PRAISE FOR An Unlikely Vineyard

"*An Unlikely Vineyard* is a rare blend of scholarship, storytelling, and poetry. Deirdre Heekin's enthralling tale of sinking roots into her land will inspire and enable anyone who ever dreamed of growing food, making wine, or bringing beauty out of the soil around them. This meditation on the cultivation of place is an elegant rallying cry in a world that too often settles for placelessness."
—Rowan Jacobsen, author, *American Terroir* and *Apples of Uncommon Character*

"Deirdre Heekin has written a colossal book here—something of a monument in its field. The author tells, in her earnest way, the entire story of establishing a biodynamic farm and orchard and garden and vineyard under improbable circumstances. *An Unlikely Vineyard* speaks to a determination and passion fueled by Deirdre's wonderful, stubborn love. The sheer level of detail may intimidate some casual readers, yet those who do read it will surely concur that it's going to become one of the Great Books of the movement."
—Terry Theise, author, *Reading Between the Wines*

"Deirdre Heekin's new book is a rural romance that's part memoir, part how-to, part coming-of-age story. As carefully thought out and set down as the neat rows of La Crescent, Blaufrankisch, and Riesling vines that populate the unlikely vineyard of the title, it's sure to be welcomed by a new generation of farmer-philosophers who will find not just inspiration but direction in its pages."
—Stephen Meuse, *America's Test Kitchen Radio*

"Not only does Deirdre Heekin take us on her own, personal path to this 'unlikely vineyard,' but she also offers us—as a *vigneronne*—the chance to understand something more universal: that authentic wine, with soul, can be crafted if one observes and takes care of one's terroir and vines. By choosing a most demanding yet most rewarding way of farming—the biodynamic way in Vermont—she is an inspiration both for farmers and for every wine lover who seeks in the taste of a grape a place, a landscape, a climate, a history."
—Pascaline Lepeltier, master sommelier, Rouge Tomate, New York City

"In *An Unlikely Vineyard,* Deirdre Heekin spins a wonderfully practical account of realizing her vision of a living farm with a table at its center. Her tale balances rural romance with the real concerns of sinking hands into dirt, of partnering with nature to bring beauty and life to her 'unlikely vineyard' in the hills of Vermont. Filled with tips and inspiration for the existing gardener, it will have armchair green thumbs ready to run off and buy a tractor."
—Christy Frank, owner, Frankly Wines, New York City

An Unlikely Vineyard

The EDUCATION *of a*
FARMER and HER QUEST
for TERROIR

Deirdre Heekin

Foreword by ALICE FEIRING

CHELSEA GREEN PUBLISHING

WHITE RIVER JUNCTION, VERMONT

Project Manager: Patricia Stone
Project Editor: Benjamin Watson
Copy Editor: Deborah Heimann
Proofreader: Laura Jorstad
Indexer: Margaret Holloway
Designer: Kimberly Glyder

Printed in the United States of America.
First printing October, 2014.
10 9 8 7 6 5 4 3 2 1   14 15 16 17 18

Our Commitment to Green Publishing
Chelsea Green sees publishing as a tool for
cultural change and ecological stewardship.
We strive to align our book manufacturing
practices with our editorial mission and to
reduce the impact of our business enterprise
on the environment. We print our books and
catalogs on chlorine-free recycled paper,
using vegetable-based inks whenever possible.
This book may cost slightly more because it
was printed on paper that contains recycled
fiber, and we hope you'll agree that it's worth
it. Chelsea Green is a member of the Green
Press Initiative (www.greenpressinitiative.
org), a nonprofit coalition of publishers,
manufacturers, and authors working to
protect the world's endangered forests and
conserve natural resources. *An Unlikely
Vineyard* was printed on paper supplied
by RR Donnelley that contains
postconsumer recycled fiber.

Library of Congress Cataloging-in-
Publication Data
Heekin, Deirdre.
 An unlikely vineyard : the education of a
farmer and her quest for terroir / Deirdre
Heekin.
    pages cm
 Includes index.
 ISBN 978-1-60358-457-9 (hardcover) —
ISBN 978-1-60358-458-6 (ebook)
 1. Viticulture—Vermont. 2. Viticulturists—
Vermont. 3. Farm life—Vermont.
 4. Terroir—Vermont. I. Title.

 SB387.76.V5H44 2014
 634.809743—dc23

                    2014018957

Chelsea Green Publishing
85 North Main Street, Suite 120
White River Junction, VT 05001
(802) 295-6300
www.chelseagreen.com

MIX
Paper from
responsible sources
FSC® C101537

Il sole sei tu. La luna sei tu. Il vento sei tu.
Antonelli Venditti, musician.

CONTENTS

Foreword

I t was a chilly winter evening in New York City. Several visiting French and Italian winemakers crammed into my very vintage, rickety, railroad apartment on the edge of Little Italy on the night before a big tasting event. When wine people converge, blind guessing games commence and I wanted to play. I chose my bottle. I plunked it inside a knee sock so no one could see the label and poured the last of it into glasses. "What do you think?" I asked.

THESE WERE PEOPLE who made the kind of wines I wrote about: no additives, as close to natural as possible, and from organic or biodynamic viticulture. They recognized what was in the bottle as one of our own and then the guesses started to pop. California! someone decided. No, another said, it's Old World, it must be . . . Burgundy? But with those tannins and those roses? Someone else interrupted, it had to be a Nebbiolo from Piemonte, not from the hills, but with that acidity, the mountains. Cherishing the moment, I leaned up against my tub, dramatically pulled off the sock as I announced, "Nebbiolo from Mars!"

They looked at me curiously, then I gave them the goods. The Vergennes Rouge we tasted was made from a grape they had never heard of called Marquette. The winery was la garagista and the wine was made by an American *vigneronne*, Deirdre Heekin. After the information I paused, looking around at the confused faces, then I pulled the punch line,

"From Vermont."

Jaws dropped and shocked laughter followed. The men and women didn't even know wine could be made in frosty Vermont, let alone that one could be so compelling and have such a sense of somewhereness.

I had been given that wine by Deirdre herself when we finally met for a glass while she was visiting New York City. I had read her book *Libation,* she had read my *The Battle for Wine and Love.* Both writers, we also had both been dancers. We believed in organic viticulture. We believed in wine as food. We believed that you could only show terroir, place, if you made yourself invisible in winemaking. At that point we barely knew each other, but she gave me a firm, intense hug full of somewhereness. But then she also gave me her wine to taste and I worried. I liked her so much, would I like the fruits of her labor as well?

I took the Vergennes home. The first taste was marked by searing acidity. By the second day, its balance was more in focus; then the rose-petal jam and riveting leathery nose kept on doing somersaults. The wine was singular. I found myself riveted, as were the winemakers in my kitchen. Her efforts had that essential quality, a sense of place.

If any wine could make a case for terroir in Vermont, Deirdre's was the one. I remember tasting the arugula she and Caleb, her husband and chef of their Osteria Pane e Salute restaurant, grew near their vines. Unlike any other I had in this country, those tender greens were so feral and delicious. Munching the leaves, I noticed they had a similar and brilliant acidity as her wine. But, even though I am now a believer in Vermont's vinous capabilities, it hadn't been too far in the past when I'd felt as mystified as those visiting winemakers in my flat.

I had been asked to preside over a wine tasting at the state's annual local cheese festival. On a hot August day I approached the bottles lined up in front of me with an arched brow. I was even more dismissive of the efforts when I later walked the vines of a well-known winery nearby. Looking at the earth, scarred by the weed killer Roundup, I asked why they didn't work organically. And, really, it was almost like asking why, out of the six hundred certified organic farms in the state, there were no certified vineyards. The growers told me the answer without flinching. Organic viticulture was too difficult in a state more famous for its spring mud season than for fermented grape juice. I asked them, even as I knew it was going to make me very unpopular, "How difficult it is only depends on commitment, doesn't it?"

"Difficult" as an obstacle never stood a chance up against Deirdre's convictions. I've seen her in action with the vines near her home; I've seen her battle with difficult vintages, refusing to alter her course; and now, having read this lyrical farming memoir, my knowledge and appreciation of her have only been reinforced. She approaches weeds in the same holistic, intelligent way that she deals with the even more pernicious Japanese beetle.

Her tales of battling the bugs were an inspiration to me. But, while I was there applauding her actions, I could just imagine others thinking, "Resort to soapy water and beneficial nematodes when a simple spritz of pyrethroids could do the job? Really?"

Really.

More difficult for me is to believe that in a state so committed to organic and unprocessed food, Deirdre is currently the sole voice for the same kind of wine. But in writing this book, she proves to all who endeavor to make true wine in climates where grapes struggle for ripeness that it is indeed possible. Others will follow. How could they not when the results are so stellar? Avoiding polemic, there are crystalline messages embedded in her prose: There are no shortcuts. There is no clear line between winemaking and viticulture. The transformation of the vine is equal to the transformation of grape into wine. But in her details—from pruning to trellising to the raccoons to the bees to the selection of cold-hardy grapes and to the passing of the seasons—I saw another transformation as well: Deirdre morphing into what the Italian call a *contadina*, a farmer, a steward of the land. The expression of her vineyard through the grape itself is nothing less than an extension of that very first hug she treated me to—sensitive, yet powerful and grounded.

ALICE FEIRING
New York City
May 2014

Table

I n the barn a long table is set for dinner. We gather here on a
chilly September night around this table, built from boards
of our old derelict garage, sitting to a very late supper of hot and
hearty soup made from the garden's vegetables: a mixture of
lovage, zucchini, tomatoes, and pearl white beans.

THE WHITE BEANS, grown on the tall trellis in the walled garden, came from
a bag of heritage Badalucco bought in a shop in the mountains of Liguria, a
bag sold for soup, tied with a pretty jute ribbon, which we brought home in
our suitcase and planted in the ground. Loaves of bread sliced on the table
and a wooden board covered by a half wheel of a local alpine cheese are the
only accompaniments. We've already feasted on links of dry *salame* and a
flatbread that my husband, Caleb, makes every fall that's studded with our
black wine grapes, rosemary, anise, and sea salt in honor of the wine harvest.

I call this the barn, though its original structure was intended as a garage,
an all-purpose building that was meant to house a car in the winter and to
store our workmanlike belongings, outdoor furniture, or overflow from our

house—pieces of a personal history that no longer fit in our living spaces but that we are both loathe to part with for sentimental reasons.

The garage is built like a carriage barn, a classic shape with a high peaked roof to accommodate a loft, a building Caleb designed with pencil and paper over the course of an autumn. It was raised in the winter, a New Year's present to ourselves, as we dismantled the old forlorn garage board by board the week before Christmas.

Inside the barn stands a makeshift buffet that's built from one of the old counters we used to have in our bakery. The galvanized metal counter that served in the early days of the bakery has been put into use elsewhere, so I've fashioned the top out of old cabinet doors. The shelves in the body of the counter used to hold flat cardboard boxes that would pop up into shape like clever origami. There were spools of vegetable-dyed string, paper napkins, wooden stir spoons, and stacks of paper bags in various sizes. Now, we store tins of nails and screws, buckets of tools, and odds and ends of building stuffs in here.

Candles march unevenly up and down the tabletop between linens, silverware, and an assortment of soup bowls from the china cupboard. Bottles of wine wend their way down the table as well, wines that lend themselves to conviviality.

Several seasons after building this new structure—a kaleidoscope of snowy and icy winters; fragrant green springs; hot, not-so-lazy summers; and colorful, bountiful, slightly melancholy autumns—the barn serves another purpose altogether: We are here to harvest and press the new wine. We sit down to share a meal together after a long day of work, warmed by the soup and flickering candles and glasses of last year's dark red vintage. Wood smoke is in the air and the barn smells of crushed, sweet fruit and wild yeast.

Our hands sticky and stained from both white and red grapes tell well the story of collecting the bunches in the field, transporting them to the barn, weighing and unloading the fruit on top of the cunning screen that Caleb built from ½ x ½-inch mesh that sits on a table covered in a red-and-white gingham oilcloth, a table exactly the same as the table at which we are sitting down to dine, both tables made from those old boards from the old garage. Over the screen, we softly roll the grapes, destemming and sorting as we go, waves of little jewel-colored berries—deep garnet, violet amethyst, celery tourmaline, or rose topaz—being pushed off the table into the big open vats. Here, friends will strip down to bare legs, wearing shorts or bathing suits despite the edge to the air, and lower themselves into the barrels to stomp and massage the fruit, gently breaking the skins to release the colored juice. Late at night, we put on vibrant music like salsa, or disco, or old jazz, and those in barrels dance and race to complete the *pigéage*.

Harvest for us now usually lasts a month or so for the heavy work, with a steady stream of friends and family joining us in shifts to sort, stomp, press, and break bread together. While the work buzzes along, we provide an unbroken ribbon of dishes born out of our gardens or larder and the farmers' fields down the road. In the morning there is a traditional *casse-croute*, a snack of simple apple cake, or toast spread with warm chocolate served with hot coffee. If we are in the field picking, we might pack a picnic of tuna and

rice salad, an herbed frittata, a long platter of cured meats, a hunk of cheese, and short, sturdy little glasses of rosé. At the farm, if we are well into bringing the fruit to fermentation, we'll sit at a long table in the garden with plates of roasted sausages with inky grapes and onions, the grill fire lending a visual warmth. There is always something to nosh on and a glass of wine close at hand. Dinner is served quite late, usually around nine or ten o'clock, and is simple like this white bean soup.

Tonight is the second night of harvest. The white grapes destined for classic dry white wine have all been pressed and funneled into the large 14-gallon glass demijohns where they will spend a significant part of their adolescent lives. The full harvest moon crests the tree line at the bottom of the meadow, and as it rises so does the cap of grape skins. We witness the whoosh of the start of fermentation as the grape must pushes to the top of the big glass jars and the magic turning of water into wine begins.

Around the table, we huddle in our thick sweaters and vests, pressing closer to our bowls of steaming soup and the small heat from the candles. While there is a definite chill in the air this night, we begin to feel warm with the hot food and glasses of bold red wine, warm with conversation and friendship. We delight in the moment of sharing food together at the table, food that was harvested right outside the doors of the barn; we delight in the suspension of all else but the imminence of the harvest. In honor of sitting down to the table in such companionship, and joined in the hope for next year's good wine, we raise our glasses together.

Landscape

🌸

We live on a small hillside farm in south-central Vermont, a piece of land that has been home to us for almost twenty years and has been the landscape for the many cycles of lives that have touched ours: births and deaths, weddings and funerals, love at first sight and lifelong romances.

THE FARM SITS UP HIGH, just below a green schist crest of the Green Mountains, believed to be some of the oldest rocks on the Earth. The farm is about 1,600 feet above sea level, having eroded over the last four hundred million years from mountains that once loomed 8,000 feet high, in a protected and conserved forest and meadow area called the Chateauguay No Town. Fifty thousand acres of relatively undeveloped land comprise the Chateauguay, and it has always been a rather remote and wild terrain. In the 1850s, desperadoes hoping to get rich quick came to the Chateauguay as part of the Bridgewater Gold Rush, but in the end no one came away with much gold. One family wrote that they mined only enough to make a wedding band. The land of the Chateauguay proved to be too unruly for easy riches.

As far as I can tell, the name Chateauguay originated in France, from a small community outside of Dieppe in Normandy. A friend recently mentioned that he had heard the name was a Frenchified version of the Indian *Sha-taw-gay*, but I have been unable to find any other reference to support this theory, either written or in the local oral history.

According to the French, in the 1600s a young Robert Lemoyne was shipped to the new world of Nouvelle France at the age of twelve and was sent to live with the Mohawks to learn their language and ways. When he was of age, he was granted a parcel of land, which he named Chateauguay after his father's village. In Quebec there is a town named Chateauguay, and a river. No one seems to know exactly why this name has been given to a dense interior land in the Delectable Mountains, a small set of hills a little more than 100 miles south in Vermont. And no one seems to know why there is a village designation of No Town, except that now there literally is no town here. Only wild bracken overtaking the remnants of cellar holes and the occasional pottery shard dug up by a curious crow.

It is here in the midst of this Chateauguay landscape that surrounds our home that the idea for this next journey reveals itself to me, the adventure that is a book about to be written. The hope of the unfolding narrative begins to come together like the quirky and varied layers of a shadow box: an assembling of scraps of images, scents, tastes, stories, memories, and elements that comprise the place we"ve begun to call our farm.

Many years ago, Caleb and I found ourselves living abroad in Italy, falling hard for a culture and her cuisine. Upon our return, we became smitten with the notion of opening a small village restaurant, fed by what the local hills and valleys could produce. That restaurant, which now exists, began as a bakery and café and has evolved over time into our little village tavern, or *osteria*. This was the beginning. As time passed, we became taken with another idea: that of growing our own gardens to provide for the osteria.

This written journey that is our farm involves the planting and tending of those vegetable gardens for the osteria, and the planting and eventual harvesting of our vineyard and orchard; it concerns the making of wine and the notion that life can be lived in both work and play, in a way that offers an honest sustenance. It's about wine and about a natural agriculture that encompasses the ideas of a complete farming landscape, philosophies like permaculture or biodynamics, and forest-edge ecology. It's about naturalistic wine, and what that really means. It's a story of the landscape cultivated for the table.

This part of our story is about learning and making mistakes, about looking to those who inspire us. This story is a portrait—in both broad and fine strokes—of our vineyard and the day-to-day workings of the farm that defines the landscape and the harvest and wine it produces. The wine being the alchemical liquid that pulls us outside of ourselves and teaches us something new: to be winegrowers who work as closely with the land as our human hands will allow, both in the field and in the wine cellar.

Modern agriculture has an affinity for language; we've seen words like *green* and *sustainable* go in and out of fashion and back again. We see farmers searching for new ways to articulate how they might responsibly husband the future by turning up the past. In wine, we use the old French word *terroir* to discuss wines that somehow speak of a sense of place. And while this word has begun to be used more widely and loosely to describe a kind of culture and sensibility, and has begun to lose some of it fashionable luster, I believe it can never go out of style or become irrelevant. There is no other way to sum up the aggregate philosophy of the true farmer. Terroir is many-faceted. The general assumption is that this word speaks only to the geology of the land in which we grow wine, or fruits, flowers, and vegetables, or livestock. But this word encompasses more than merely schist and clay, sand and chalk. Terroir is about mud and stones, but it is also about the varietal nature of the plants or animals that grow in or on this land, the microclimate of a hillside or plain, and the personality of those who do the tending. It represents the six sides of the honeycomb: geology, variety, geography, climate, social culture, and the human hand. Another winegrower I know says that what is poured into the glass is a liquid landscape painting of the 365 days of a certain year.

This is my hope on our farm: to capture the four seasons of each year in the bottle, a liquid portrait of our landscape and its history. This is my quest.

Proserpina is the name of the Roman goddess of the seasons: She represented the springtime growth of crops, which led to the eventual harvest of the autumn, and the whole cycle of life, death, and renewal. She was the daughter of Ceres, goddess of agriculture, and of Jupiter, god of the sky and weather. Her counterpart in Greek mythology is Persephone, who lived in the underworld during the winter months and emerged aboveground, just like the plants she renews, with the first signs of spring.

When I began compiling this book, I was reading an old text on agriculture by Virgil, a text that is closely linked with the ancient agricultural practices that form much of the basis of what is now known as biodynamic farming. Proserpina is mentioned in the *Georgics*, as is her mother, Ceres. Together they form a powerful duo that oversees the world of nature and the human cultivation of the farm. Alongside Jupiter, who rules the skies, I see them form a triumvirate of terroir. Proserpina's name is believed to come from the Latinized version of Persephone, the word *proserpere* meaning "to emerge or creep forward." This is often associated with the growth of grain because of her mother Ceres's connection to grains, but I've often thought it would be more accurate to connect the definition with the grape vine. In French, a grape vine is called a *plante grimpante*, a plant that climbs or creeps upward. In Italian, they are considered *rampicante*, twining enthusiastically toward the sky. The kinship between the seasons of Proserpina and the life of the vine, the life of a garden, or of a farm, appeals to me, and it only makes sense that the mythological goddess of the growing season would preside alongside her mother and father over the cycles of the year. These gods, as personifications of nature, represent the real painters and storytellers of that liquid landscape in our glasses.

Whenever I speak or write of our farm, the farm always asserts itself, as if it were spelled in capital letters, because we never imagined that the terrain we've come to call home would grow into a working and edible landscape. I never imagined I would be a farmer, let alone a farmer with a quest. To be fair, I never even imagined owning a restaurant, or growing wine. I've written elsewhere that at the center of this farm is an unlikely vineyard, a geography so far north it seemed madness to try to grow wine. But I've also written elsewhere that I am an unlikely winegrower. Most of my stories start with this notion, "I never imagined . . . ," or "Twenty years ago, I would never have guessed . . . " I seem to repeat myself, but they are true lines, and I make the most sense of deciphering how I spend my days by stepping away either physically or metaphorically, and relaying the adventure with words and images that I shape around each other.

Several years ago, we began to turn our 8 acres of land into a farm, returning it to an earlier point in its history. Our parcel was once part of a much larger subsistence farm, and by subsistence I mean the old-fashioned notion of what we now call "sustainable." The land was used as a dairy farm, with a large cow barn (which has now been turned into a second home), rough stone walls, open pasture, forested hills for logging, a farmhouse garden that grew vegetables for eating during the season and putting up for the winter, plum and apple trees, and naturally growing wild edibles in the fields and along the streambed dissecting the land. That old farm fed the family that lived on it plus the land created a living for them in the world outside the farm. But once dairy farms began to slide into economic complications, and the old farmer had gotten too old to work the land, the property became divided into smaller parcels, and houses were built as single-family dwellings or summer camps. Our house was constructed as a summer camp that sheltered the old farmer and his wife, and the originally divided parcel became even smaller over the years, pieces sold off to keep the old farmer healthy and his head above water. It is an old story, a melancholy story, a true story.

By the time we came along, the house had changed hands once. The meadow had been left to return to the wild, full of ferns, goldenrod, and the march of young poplar. The once locally renowned wildflower garden of the lady farmer had been pulled up or gone to seed; only afterthoughts were left of spring daffodils, a hedge of wild rose, and an exhausted asparagus bed. A little shed remained, painted dark red with white trim, techni-

cally a potting shed that had once been used to house two baby lambs for a season, to protect them from the prowling coyote of the Chateauguay, that wild forest above and around the old farm meadows.

When we arrived on the land, we did not know how to farm. We barely knew how to garden. Our first gardens were lessons in humility. Garden beds were rampant with bolted chicories and threaded through with grass and herbs, invasive things that spread—a place for our cat to hide and catch mice. We never properly prepared the soil; we didn't know how to tame the field that surrounded our gardens and wanted to overtake them. Those first years were misguided and sheer fights against nature. And everyone knows the outcome to that tale.

The next two garden spaces we built did improve with raised beds and compost. Beds that we originally thought might have roses and ornamental fruit trees soon were growing lettuces, carrots, onions, beans, herbs, beets, celery, fennel, cabbages, radishes, and tomatoes. Before we knew it, we were harvesting produce to take to our small osteria where we cooked and served simple dishes and honest wines, and the small orchard we had planted began bearing fruit for tarts, cakes, and sauces. By the time we planted the first vines and the first walnut trees, we suspected that what we had on our hands

was a farm, but it took us a little while before we were brave enough to call it by that name. We hemmed and hawed around the words *gardens, vineyard, orchard*. We were cook and sommelier, dishwashers and writers, floor scrubbers and servers. By the time we had tucked in our first wine and cider harvest for the winter, and by the time we were pruning the apple trees in two feet of snow, we realized we had to come clean. We had a vineyard. We were growing wine. We owned a farm. A very small farm, but a producing farm. And we had become farmers.

A Note on the Farming

Just as recipe books often have a chapter called "A Note on the Recipe" to give the reader an orientation on how to approach the recipes and in what cooking style, a book on farming should orient the reader to the experience, the kind of thinking and the kind of farming shaping the methods and reasons behind the agriculture and the farmer.

THERE ARE AS MANY different methods of farming as there are farmers. A farmer's philosophy develops over time and changes with experience and knowledge. There is always a learning curve, even when farming is in your blood and it has been passed down to you over generations. And some farmers are made from the most inauspicious stock, without a speck of farming heritage in the mix (though I suspect, somewhere along the line, we all come from farmers). And sometimes true farmers resist farming. They try all manner of other occupations and arts, but somehow are never really satisfied until they find their hands in the dirt.

Farming History, Ours

While it sometimes seems strange to both my husband, Caleb, and me that we choose to spend a good chunk of our time working with plants, reading books on horticulture, gazing at seed catalogs, and arguing the merits of the best gauge of wire for a trellising project or possible solutions to pesky drainage issues, we both don't have to look too far back in our own ancestry to find people who worked the land.

CALEB'S MOTHER'S FAMILY FARMED the tough granite soil of New Hampshire, and he grew up with his mother's flower and vegetable gardens here in Vermont. He was often assigned the task of weeding the beds, and at a young age this was a horrible fate; by his own admission he much preferred the idea of riding his bike as fast as possible down their steep hill or playing at some other far more interesting game. His father's family, most immediately, comprises lawyers, artists, and intellectuals, but only a few generations back they too farmed land in southern Vermont.

My own most immediate history was with livestock: horse farming. My grandmother's family trained and stabled horses, and my sisters and I had ponies to care for at a very early age. My grandmother's family had also once been silk farmers in Switzerland before coming to this country in the late 1800s, and my paternal grandfather's family were subsistence farmers in Ireland at the time of the Potato Famine. On my mother's side, farming went back only one generation to the Swedes who settled in this country, homesteading north of Chicago.

In southern Indiana, where I grew up, we were surrounded by agriculture, and on every scale. Fields of corn, soybeans, sunflowers, and prized Posey County melons filled the open land outside the large town we called home. We lived on the edge of that farmland with just enough space to pasture our growing farm of horses (my parents unknowingly bought two mares that were awaiting foals—a two-for-one deal), and my mother had planted extensive flower gardens. My parents had tried their hand at vegetable gardening when I was very small, but because there was so much superior produce near at hand, we devoted that area to a small paddock for the newly born colt and filly.

Caleb always thought he didn't care much for the physical act of gardening, whereas cooking was becoming a slow simmering passion for him, and I never considered that a garden would figure in my daily life. Our experience living and working in Italy when we were young and newly married changed all that. It led us to the appreciation of food and wine first, to all things that unfold at the table. Sitting at the table led us to standing in the kitchen to learn how to prepare all those dishes that we so enjoyed on the plate. We learned from older cooks who taught us to work by our senses, through touch, taste, and scent. Recipes shared were spoken, and we either had to commit them to physical memory or write them down so very quickly on the back of a paper napkin or grocery receipt in order to capture them. We learned to dissect recipes simply by the experience of eating them.

Cooking food, along with the desire to eat well, led very naturally to the ingredients. In Italy, these are called "prime materials," and great importance is attached to them. While living abroad, we grew accustomed to being able to procure everything we needed within a very limited radius and to watching our friends and neighbors working their own small allotments, picking lettuces and tomatoes or leaning an orchard ladder against a tall tree bursting with ripe cherries. By the time we moved back stateside, we wanted the same. This was before the proliferation of small local farms and farmer's markets. So, in our first summer back on this side of the sea, we planted our first garden.

We lived in an apartment on the main street of our village, and borrowed a garden plot from a friend who had a house in the countryside. Our allotted space measured about 12 feet by 12 feet of rough ground that got tilled at the beginning of June. It sloped slightly downhill and became a battleground for chickweed, orchard grass, and bedstraw. We bought a garden fork, a shovel, a hoe, and a couple of weeding claws. We kept our site simple. We planted mostly tomatoes and herbs, a lot of basil and zucchini. We planted green tomatillos to make a salsa we'd become enamored of to serve with ground lamb tacos and chèvre—a recipe that had quickly become a staple on our in-house menus.

We planted some annual flowers like cosmos and snapdragons for cutting. Caleb made a tomato trellis out of reclaimed scraps of wood to hold what proved to be a prolific crop. There was so much satisfaction to be had in that first little garden, to go care for it in the soft, humid evening or on a hot, sunny Saturday morning, or a peaceful, cloud-covered Sunday.

We gardened by the seat of our very dirty pants. We knew enough to keep the space around the vegetables clear of weeds. We'd heard that basil loves tomatoes in the garden as well as on the table, so we planted the leafy herb in the shade of the sprawling vined plants. We knew about compost only because Caleb had grown up with it, but we didn't amend the clayey soil in our plot with anything. We harvested beautiful little Sungold cherry tomatoes that season—a bounty I'm not sure we've ever seen since. We made jars and jars of a velvety golden sauce, and certainly had plenty left over to eat fresh in a homely dish of a short pasta with the yellow tomatoes, aged mozzarella, and those sturdy basil leaves.

We continued with that garden plot for several years until we moved to open our bakery and café, our first foray into creating the small restaurant we had set our hearts on. We didn't vary much in our gardening skills and didn't do much to study how to improve our efforts. The vegetables grew well enough, and we were happy to have any kind of bounty. When we moved, our new apartment had enough yard around the old Victorian house that we were able to keep our garden close by. Given that it was the first year of running our own shop, I still can't believe we found the time to rent a roto-tiller and prepare the ground, let alone plant anything and care for it. But we did. I admit I don't remember much about that garden, except the snapdragons that I planted there. I remember always planting snapdragons. And that the tomato trellis we had moved with had rotted and was no longer useful.

It was only a year and a half before we bought our house in the Chateau-guay, and in our first summer there we tilled the first long beds of a *potager* (kitchen garden) in the middle of our overgrown meadow. Still, we knew nothing. We blithely planted herbs and all that wild mint taken from my mother's garden, as well as a host of different radicchio and chicory seeds we'd brought home from our travels in Italy. Caleb was determined to learn how to grow radicchio. Somehow, that first year, those were the only things we planted: herbs and radicchio. Work at the bakery intensified, much-needed house projects beckoned, and the garden very quickly became out of control. Everything we had done to prepare those beds was not enough, and the work became overwhelming. We could not stop the advance of deeply and broadly matted grasses, goldenrod, and sensitive fern. By autumn, I distinctly remember sitting in front of those beds in a worn Adirondack chair that we had found tossed in the woods at the bottom of a neighbor's field, feeling the weak sun on my face and the exhaustion in my bones, with the garden a tangle of overgrown, tough-stemmed radicchio riotously mingling with every weed our soil would support. A light breeze was blowing, chasing our cat up and down the field. The mess of the garden was crushing. We knew we could no longer garden as we had been. The scale, the soil, the landscape were all different, and we were distinctly wading in tall meadow grasses, over our heads.

We set out to learn to garden just like we learned to cook. We asked advice. We visited gardens and talked to their gardeners. We devoured pictures of gardens, learning something from what we saw there. When we traveled, we spied on allotments and made new friends when we saw someone plant-ing potatoes or hoeing manure into their plot. In the winters, to brighten the quiet of snowy evenings, I began to read narratives of gardens—tales of small, imperfect gardens by a river or the resurrection of formal gardens at a great house. Caleb's grandmother, who no longer had a garden, sent us old, dog-eared books from her library about companion planting and heirloom vegetables. She had belonged to a group in her village called The Plain-Dirt Gardeners. We learned from our blunders and read the wisdom of others.

Caleb's mother too shared her knowledge and loaned us books from her own early years of gardening. When we went to visit, Caleb and I would comb through the floor-to-ceiling bookshelves for gardening manuals. The books we found were certain treasure.

While all of this was happening, my study of wine was also escalating. The bakery became a small restaurant, the osteria, and I devoted myself to the front of house, making an inviting place for people to come dine and managing the constantly growing and changing wine list. Initially, I focused on the varietals, the regions, and the work done in the winemaker's cellar to know something of the stories of the wines I was serving. My mission was to learn the regional symbiosis between lesser-known varietals of a certain locale and how they wove inextricably with the food that was grown and the recipes that were developed from the ingredients grown in a particular place.

Through my mentors in wine, and the education of my palate, I began to understand the notion that good wine is made in the vineyard, not the cellar. So on our yearly pilgrimages back to Italy to reconnect with the sensibility, food, and landscapes that had so inspired us, our visits to wineries became about visits out into the vineyard first, followed by a tour of the winery and tasting from barrel and bottle alike thereafter.

In wine, I believe terroir is key, and soil is a major part of any discussion with a winegrower. Soil may be a cornerstone consideration, but it is hardly the only point to the story. There is also the microclimate, the varieties of grapes, the geography of the land, and the winemaker herself. Gardens also respond to the same aspects of soil, microclimate, variety, and the human hand. All these pieces interact, fit together in some fashion, and must be taken into account.

Why Organic? Why Biodynamic?

By the time our gardens had grown enough to slip into the guise of a small farm, we knew that we would be pursuing our work organically. Our teachers worked organically, our family mentors worked organically, and we fully believed in the imperative of properly stewarding the land without the use of manmade chemicals. We had learned enough to know that the viscious cycle of chemical agriculture eventually kills the land and the finely wrought ecology that sustains it.

CHEMICAL AGRICULTURE IS A SEDUCTIVE siren. In a particularly rainy season, where rain had fallen every day, overwhelming our soils with flooding water and compromising our plants, I saw other growers' vineyards managed more conventionally. While we practiced triage on vines riddled with black rot in one of the blocks of our home vineyard, their vines looked pristine, the leaves lush and green, the fruit unmarred and fat. In addition to some traditional chemical sprays, they had excellent drainage, and the rows felt relatively dry beneath the feet, not spongy or mired in mud. In comparison, it was demoralizing to see the black and brown speckles on our leaves and fruit, and as it continued to rain, those plants affected the most succumbed to the die-off of new growth that was inevitable.

For a moment, there was a part of me that understood those farmers who have tried new chemical products offered by their local agricultural extensions. Their plants become blemish-free, the fruit succulent and beckoning, the eventual harvest high in yields. I understand daily that the word *sustainable* must also embrace economics as well as the environment. To lose a crop can kill a farm. In some cases, a chemical tool can save a farm from disaster. But this vision of perfect health and beauty created by the constant application of synthetic inputs is a mirage at best and only on the surface, literally. How long can the soil and the plants sustain such brilliance? As it turns out, not long at all. The roots of these chemically maintained plants stay in the topsoil, spreading themselves out horizontally and lazily. The conventional (i.e., chemical) spray program will falter in three, five, or ten years' time, and other new products will need to be used as the original chemicals become ineffective. In chemically treated vineyards, many farmers budget the replanting of vines every twenty years, knowing the vines cannot survive longer than this. They cannot sustain themselves. I think of organic old-vine vineyards in France, Italy, even in California, that are approaching a hundred years in age. Because they are so old, these vines may not offer high yields, but the wine they make tells a long and varied story, a history. Yes, I had one section of my vineyard that was ailing in that imperfect season, a block that I failed to treat preemptively and organically quickly enough when the season

began, and, yes, those vines looked rather desolate in their continued and stymied efforts to bring forth new growth. All the work we did focused on controlling the damage that had already been done for the season.

In that same season, I looked to my block of white grapes. They, like the other growers' vines, had beautiful, vibrant leaves and luxuriant growth; the fruit in the bunches had begun to establish itself. Here, the fungi had been kept at bay, and my momentary mental dalliance with conventional seduction evaporated. The mirage was gone. I realized that I am content to know everything that goes into and on these plants, and that our efforts are the result of a balanced and natural agriculture that begins from the soil upward. These vines work in tandem with their land, and the wine we produce from them will be an expression of that hard-won balance achieved in any particular year. While our red fruit harvest suffered in that particular vintage, we will not be enticed away from our belief that working *with* nature is essential and that fighting *against* nature is foolhardy and born out of fear. We have rather more Arcadian dreams.

We first became interested in biodynamic farming through wine. The wines that I gravitate toward for the wine list at the osteria always seem to be made by producers who are certainly organic and, more often than not, quietly biodynamic. There was a period of time, even in Europe, when biodynamic farming was scoffed at, and those practitioners who produced their crops that way preferred to keep their growing practices to themselves. When a few prominent wine stars revealed their hand, like Gravner in the Friuli or Domaine Romanée Conti in Burgundy, biodynamics found the spotlight. Other producers came forward, and before anyone knew it an organization called The Return to Terroir had been formed by the grand Loire producer Nicolas Joly, a torchbearer and teacher of biodynamic thought in the vineyard. Soon, tastings featuring only biodynamic wines and lectures on biodynamic wine were organized. These shifts in consciousness always take time to brew and take hold, but it seemed as if, almost overnight, everyone in wine was talking about biodynamics in the vineyard. More and more, strictly organic producers were transitioning to biodynamics. Even big-name and philosophical producers here in the United States, like Randall Grahm of Bonny Doon Vineyards in California, converted and took up the mantle.

The timing has been right for these wines, and those who speak eloquently and vehemently about the need to understand the principles and the positive effects on the wines and in the vineyard are being heard. But what really is it about biodynamics that has been so attractive to winegrowers who want to grow honest and memorable wine?

In the world of wine, and more and more now in food production, biodynamics separates itself because it considers flavor. There is often a disconnect between the growing of food and the cooking of food. Gardeners and farmers don't necessarily aim for producing heightened flavors or ingredients that speak of a place. Organically practicing farmers want to take care of their land, plants, and animals responsibly, provide the best and healthiest nutrition to them and, through this food, to those who purchase the vegetables or meats in turn. These farmers are most concerned with the ins and outs of the growing season. It hasn't been until recently that professional chefs and cooks have gotten into gardening or into thinking specifically about how they are sourcing their food and the type of farming practices used. Professional cooks have to consider another level as well: How does it taste? Will it satisfy the many aspects of the palate? It is not just about the moral imperative.

Conversely, wine producers have for a long time kept the concern of taste front and center. Wine is all about scent, taste, and impression. The language and focus around wine discusses wine like a story: There is a beginning, a middle, and an end. There is an arc of taste. There are high notes and low notes. Depth, breadth, verticality, circularity. These structures are filled in with fruits and nuts, savory herbs and minerals, dirt and animal magnetism. And many winemakers are the growers as well, or, if they buy the fruit, are still very involved in the growing program. There is no disconnect between field and cellar as there can be between field and kitchen.

Biodynamic practices focus on how to be organic in terms of farming principles and how to elicit the best flavors and most intoxicating experience. Winegrowers who want to reflect the heart of their landscape through their wine would naturally be interested in pursuing a growing program that concentrates their efforts into good cultivation practices that in turn produce wines of great character.

When I came to the conclusion that I wanted to grow wine on my small parcel here in Vermont, I wanted to do it biodynamically. I had tasted the difference in wines grown this way. There are no exact answers as to why the brightness and personality of the flavors are so pure, but in the best wines that exemplify this, the purity shimmers.

It is one thing to decide you want to be a biodynamic winegrower, and another to do it. It can be incredibly labor-intensive. Some producers blend biodynamic sprays with more typical organic spray programs using sulfur and copper. Other producers use only plant teas and compost. Most practitioners admit that it takes a long time to understand your land well enough to manage the vineyard completely without the minerals copper and sulfur.

When we first planted the vineyard, I immediately began working biodynamically according to the calendar and using various plant teas to keep the vines healthy. By the second year, I was spraying some kind of tea nearly every day. The vineyard was small enough that I had the time to do this before I went into work at the osteria and, on farm days, before I had to devote time to other projects. We had quite a bit of rain that year, and by August we noticed a ring of downy and powdery mildews at the edge of the vineyard. In and around the vines and down the rows, however, there was no mark of any fungus. It was clear that whatever I had been doing had worked to fend off any pathogens. That experience alone proved to me that it is worth pursuing the methodology. I became a believer in the process, and we have continued to farm under a biodynamic umbrella, adjusting our practice to the season, ever since that time.

TODAY WE TAKE THE WORD *organic* somewhat for granted. As consumers, we think we know what it means. We make assumptions about the stringency of rules and regulations that govern certified organic or biodynamic farming. We know that a farmer isn't allowed to put the word *organic* on a food label if he isn't certified, even though he may follow the original tenets and aspirations of organic agriculture but can't afford to pay for the certification process or doesn't believe that a farmer should have to pay for working naturally. This begs the question: Why should farmers who work naturally be penalized by having to pay for a seal of approval? Perhaps chemical farmers should pay for their choice to use non-natural materials? Their farming will be easier and their yields higher, though it will be just a matter of time before the soil will be depleted of nutrients and a rich diversity of microbial life, and a perennial crop like grape vines will need to be replanted.

UNDERSTANDING
PROGRESSIVE FARMING

ORGANIC

Most people think that old-fashioned, natural farming fell by the wayside after World War II when all the excess munitions left over from the war needed to be used (explosives are made with nitrogen compounds) and got redirected to agricultural applications. In truth, chemical farming began long before that time. As early as 1840, German organic chemist Justus von Liebig became known as the "father of the fertilizer industry" because he found that nitrogen, a major plant nutrient, could be applied to plant roots in the form of ammonia. He promoted substituting chemical fertilizers for natural manures, and many farmers embraced this "clean" and easy solution, having no idea what the long-term effects might be.

Bringing farming into the lab rather than keeping it on the farm gave birth to a kind of collective agitation to examine the differences between the new chemical farming and the traditional old ways. Sir Albert Howard, often described as "the father of organic farming," was one of the most vocal proponents of traditional farming practices.

His experiences in India—learning from local farmers about the Indore process of composting, which has become the basis of our modern efforts at soil improvement—drove him to highlight the relationship between soil, plant health, and, in turn, the health of the people eating those plants. This is the core of our modern organic farming thought: Healthy microbial activity in the soil is the goal of every organic farmer. In organics, it all begins with dirt.

BIODYNAMIC

Biodynamic farming (from the Greek words *bios*, "life," and *dynamikos*, "powerful") is usually associated with or accredited to the anthroposophist philosopher Rudolf Steiner, but like most natural farming methods it actually existed long before Steiner ever arrived on the scene. In 1924, at around the same time that Sir Albert Howard was writing about Indian compost, Steiner gave a series of eight lectures on agriculture at Schloss Koberwitz in Silesia, Germany. His lectures, the first to have been given on the "concept" of organic agriculture, were held in response to a request by local farmers who had noticed degraded soil conditions and a deterioration in the health and quality of crops and livestock resulting from the use of chemical fertilizers. Steiner collected stories and information on old farming practices going back to Greek, Roman, and even Egyptian agricultural beliefs and methods, and codified these ideas into a generalized program that an early twentieth-century farmer could follow in order to bring his farm back into balance with nature.

Biodynamics, as presented by Steiner, has several points in common with organic agriculture: Of paramount importance is the use of livestock manures to sustain and aid plant growth—maintaining and improving soil quality and the health and well-being of crops and animals. Cover crops, green manures, and crop rotations are used, and biodynamic farms foster biodiversity by encouraging a polyculture of plants, insects, birds, and other animal life. All these methods work in relation with one another to enhance the natural cycles and biological activity of the soil.

But Steiner went one step further. He spoke of how the health of soil, plants, and animals depends on bringing nature into connection again with the creative forces that shape it. The practical methods he gave for treating soil, manure, and compost, coupled with the use of

plant medicines in the form of various plant teas, were intended above all to serve the purpose of reanimating the natural forces, which in nature and in agriculture were seen to be on the wane and at risk. In Steiner's lectures his goal was to reconnect the forces of the earth with the forces of the sky, to remind us that our living planet is not immune to the trajectories of the moon, the sun, and the movement of the other planets through the heavens. The crux of Steiner's argument centered on the realization that, if the phases of the moon control things as elemental as ocean tides and the female cycles of reproduction, then surely plant life too will be affected by such strong influences. And, if we observe and pay attention to the responses of soil, plants, and animals to the energetic powers of the planets, we will become that much better connected to our growing techniques.

NATURAL FARMING

In 1938 Japan, Masanobu Fukuoka had an inspiration that changed the course of his life. It became clear to him that to try to improve upon nature is folly. He decided to quit his job as an agricultural chemist and to return to his family farm to create a concrete example of his newfound understanding and appreciation of nature by applying it to agriculture.

Fukuoka worked to develop a system of natural farming, or what has been called "do-nothing" farming, despite its being very labor-intensive. Natural farming, or ecological farming, is related to organic and biodynamic principles, yet different. His system is based on the natural ways of living organisms that shape a particular landscape and ecosystem. In ecology, mature ecosystems are incredibly stable with high productivity and diversity. An old-growth forest is a good example of this. In Fukuoka's vision of natural farming, his effort was to mimic these virtues, creating a comparable agricultural ecosystem. His methods try to minimize labor and adopt, as closely as practical, nature's production of food.

Fukuoka's philosophy, presented in his seminal book *The One-Straw Revolution*, was in direct response to the land he farmed and the circumstances surrounding it. And while each farmer in every location may not be able to farm exactly according to Fukuoka's principles, his teachings force the farmer to observe and to farm in response to what she sees, touches, tastes, and smells within the natural ecosystem of her specific farm.

PERMACULTURE

Ecology-based farming philosophy has long roots stretching back into history. The first gatherers learned how to cultivate from observing the natural world around them. In 1929, only five years after Steiner spoke at Koberwitz, American Joseph Russell Smith wrote a book called *Tree Crops: A Permanent Agriculture*, a work in which he summed up his long experience experimenting with farming fruits and nuts. Smith saw the world as an integrated whole and recommended the use of mixed plantings of trees with other crops grown underneath. This book inspired many people who were intent on making agriculture sustainable.

The definition of permanent agriculture as a farming method that can be sustained indefinitely was more formally introduced by the farmer-philosophers Bill Mollison and David Holmgren, who collaborated on the book *Permaculture One*, inspired by the issues of water collection and consumption in Australia. After the success of that book, Bill Mollison went on to establish The Permaculture Institute in Tasmania. Meanwhile, half a world away, Austrian farmer and permaculture advocate Sepp Holzer began working on his own farm in this way—his was one of the first modern applications of permaculture concepts as a complete farming system.

Permaculture designers look to patterns that occur naturally and try to understand how they interrelate in a certain landscape, imitating these patterns and integrating them into a design that will most benefit the people and the land. Taken from the study of ecological systems and preindustrial land use, the philosophy draws from several disciplines: organic farming, agroforestry, integrated farming, natural building, rainwater harvesting, and applied ecology. It has become a branch of ecological design that encourages sustainable architecture and independent farming on a small scale and is based on natural ecosystems.

Rebel farmers and philosophers, each and every one—that is the common thread connecting these creative and innovative people involved in these postindustrial models of agriculture. They look back to ancient farming methods and look forward to how to shape them according to the needs of their landscapes, animals, and plants. They look to the natural environments surrounding them for inspiration and understanding.

We hear on one hand that certified organic assures us that our food is clean; on the other we hear that the guidelines have been softened so much by lobbyists and politicians that we can't entirely trust all the products that have been certified as "safe" to use in agriculture. We see that companies like Monsanto, the GMO seed giant, label themselves as "sustainable" and "green," and these words orbit around the notion of organic and processes that are environmentally safe. In wine, certified organic viticulture can be rendered null and void by the very nonorganic chemical work that may go on in the cellar, including the commonplace overuse of sulfites. More than seventy additives are allowed by the USDA in wine, and there is a movement of responsible winegrowers requesting that the USDA require wine labels to list their ingredients for true transparency. But many wineries have opposed this kind of labeling and have blocked the bill. Producers who strongly believe in transparency often list the ingredients on the label anyway, even though this is not mandated. Strangely the labeling regulations are shifting in a way that complicates the inclusion of this kind of exact and transparent information on the back of a wine bottle.

Even though the cooptation of the word *organic* as a marketing tool seems to have lessened its meaning to some extent, we remain hopeful and encouraged. Just when it appears that the organic fight might be losing ground, a new generation of young farmers is taking up the cause. Here in Vermont alone, the proliferation of organic community supported agriculture (CSA) farms has grown significantly even in the last five years, and continues to grow as young farmers come to Vermont to learn and share.

How we've chosen to farm, how we've chosen to design our landscape, and how we live in that landscape are informed by a myriad of natural gardening and farming methods. The core concepts are interwoven strands of philosophies that we have studied and implemented: organic, biodynamic, natural, and permaculture threads. However, we look not only to the collected methods of these agricultures, but also to our own set of mentors who affect us even more immediately and tangibly.

I think of a lunch we had one mercurial spring day at a vineyard called Sanguineto, on the outskirts of the village of Montepulciano in central Italy. To talk of Sanguineto and her wines is to talk of their farmer, Dora. The land, the vines, the mood there—is Dora. In her seventies, she is a force of nature, or a kind of reckoning—you choose. She wears a hunting cap, a wry smile, and a long ponytail tucked away. I think of her as a modern re-creation of the goddess Diana. As head of the regional Tuscan hunting association, she hunts and butchers her own meat; she cooks her own food; she grows her

own wine. She is both domestic, grounded in the daily routines of life at Sanguineto, and philosophical in her thinking outside of tradition. At the same time, she is the embodiment of tradition and holds dear her family's own agricultural legacy.

Between the risotto and the roast chicken that they raised right there (an old hen that Dora dispatched and roasted in olive oil, seasoned with nothing more than salt and pepper—tender and refined on the palate, one of the best chickens we've ever had), we talked of pruning, tradition, nature, and honesty in the vineyard and the cellar. Dora's philosophy has informed my own, the notion that wine is made in the vineyard, and in one cellar it ferments, while in another cellar it ages. The vineyard is completely worked by Dora: She is pruner, she is tiller, she is harvester. At that lunch that day, she leaned in to me as if to tell me a secret, and of all the things I have learned from her, this may be the most important: *Tu devi essere controcorrente.* "You must be against the current."

I think of all our mentors. Nicoletta, who eschewed an intellectual life in the city to live as a simple farmer among her vines. Werner, who like his fellow Austrian Sepp Holzer has been punished for his vineyards that aren't pruned; the fire set to these vines by an angry and envious neighbor put out before there was too much damage. Emmanuel in Burgundy, who has also been threatened and attacked by those who cannot understand the forward and natural momentum of his agriculture. Randall, a winemaker-philoso-

pher if there ever was one, who experiments and writes and grows wine pushing the boundaries of accepted viticultural thought and action from organic to conventional camps alike. It is no surprise that one of his wines is called Contra—from a hundred-year-old vineyard in Contra Costa County in California not typically known for great wine—as if the making of this old-vine blend from vineyards in an unlikely place is pressing hard against the established norms. They are all rebel farmers, renegades, visionaries. They are all *controcorrente*.

So, this is how we approach the work here on our 8 acres in an alpine meadow in Vermont. We have taken in and assimilated something from each of these farmers and each of these progressive schools of thought. And while all the approaches are unique, as they should be, they all hold basic values that are the same, most significantly the act of engaging your senses in the work of farming. Time and again, my own teachers have said to me that there are no pat or prescriptive answers or methods. Your own landscape should and will define and design what you do. Look and listen to what your land is telling you. Pay attention to its natural tendencies. Employ the precepts you have learned that are appropriate and applicable to your environment and your land's needs.

This assemblage of practices and the crafting of our own approach creates a new and singular way of farming, individual to us as the farmers and unique to our place—our own expression and understanding of a new synthesis of our version of a whole-farm agriculture.

Vine Yard

On the day you come to visit at the farm, the first thing I will want to show you is one of the oldest grape vines on the property. Broad as a cedar post, the vine starts down somewhere in the raspberry thicket threaded with virgin's bower, incongruously also known as old man's beard, and runs along the bank of the brook that lines one side of our property. It wends its way up into the trees, as grape vines are naturally wont to do, and sends out magnificent tendrils, arms, spirals of vine. In a good year it is heavy with fruit that feeds the birds nesting on our hill.

THIS VINE CONVINCED US that our land was proper for a vineyard. All along the brook other grape vines twine, some larger and older than others, and in the neighbor's field, a host of vines have dressed four tall pines in a mantle of spring green leaves. Here, they are prolific.

Northern New England has been covered with vines for hundreds, if not thousands, of years. Back in AD 1000, the Scandinavian explorer Leif Ericson landed on our shores and christened our northern clime Vine Land because of the profusion of native grapes he found growing there.

Some will tell you that apple orchards are a more appropriate natural partner to our Vermont landscape; we too once thought the same. And while we love apples and orchards and all they have to offer (it is why our house sits in the middle of an orchard), apple trees are not indigenous to our American land. They have been here for a long time, at least since the Puritans came to conquer Plymouth Rock, and they grow exceedingly well here, but they are not originally of this place. Grape vine is.

Once you see one of these old grape vines climbing out of a riverbank, you begin to see them everywhere. Even along the busier commercial roads, guardrails are obscured with wild grape vines in the height of summer. You see them wrapped around trees, you see them running along farm fence lines. Once you know they are here, and part of our indigenous flora, you might even see them in your dreams.

American Dreams

Those mammoth old native *Vitis riparia* vines twining around the edge of our meadow made a profound suggestion, but it took a while for us to get the hint. We began reclaiming our once-farmed land with flower gardens, vegetable gardens, and then an orchard. A vineyard was very far from the vision we had pursued in designing our landscape in relationship to the natural countryside. It wasn't until a combination of

events and circumstances came together in that flash of lightning, the quintessential *coup de foudre* moment, that the vineyard presented itself. Of course, the lightning striking is an event
seen in hindsight; I'm sure we were working up to this idea of a
vineyard for some time, a niggling sensation in the background
trying to insert itself into the grand scheme, not unlike those
unruly grape vines climbing the trees of our hedges, showcasing
the fruit of their labors.

I REMEMBER WELL STOPPING on a hot Vermont afternoon in the Champlain Valley on the way back from a visit with family in Lake Placid, New York. We pulled up at Lincoln Peak Vineyard, at the then small tasting room with a small garden of native wildflowers in front. Lincoln Peak is a winery specializing in growing wine from cold-hardy hybrid grapes bred specifically for thriving in our kind of climate. A friend who lived in the area had told us about the wines being grown there, how they surprised him by their authenticity and their resemblance to "real wine." And it was true, they weren't foxy tasting (a derogatory term used to denote a strong, animalistic element to wines made from native American stock and hybrids), nor did they have overtly sweet expressions. They had balance, they had fruit, they had savory notes. They did not have that potential "lesser sibling" quality about them that has often plagued hybrid grapes made into wine. They stood up all on their own.

Despite the clear blue sky, in the heat and humidity of July, lightning did strike in that moment, and all of sudden we knew we too could grow wine on our little handkerchief of land, and that this was the piece of our farming notions that had been missing. We had found the keystone. I've only had one other experience like it—that was meeting my husband. That little bit of fulmination resulted in living abroad, returning stateside, opening a bakery,

then a restaurant, and buying a derelict house on an overgrown corner of an old farm. That day in the Champlain Valley, we left with one hundred grape vines in the back our car.

The drive home was full of inspiration. Planting a vineyard spawned ideas of meals prepared in our tiny kitchen or over an open fire and served under a grape arbor, of a small farm that was cultivated like a large garden with each element corresponding and feeding all the others. We could see a small cantina open for wine tasting, picnics in the vineyard itself, or under the oldest apple tree laden with its golden and green fruit. Some of the ideas were not fully formed; they were just glimpses into something larger than we were. Images of cutting a head of escarole from a raised bed, a basket full of newly dug carrots, full bunches of dark-skinned grapes warm and fragrant on a wooden table, or—hackneyed as the image may seem—two hands in the dirt.

At the center of this imagined and very diversified farm comprising of raised beds, orchard, bee skeps, rose garden, berry patch, and trellised vines stands a long table with people seated. There is animated conversation; there is wine; there is abundant food. In the spring, the table is set amid white blooming plum trees that smell of sweet almonds. In summer, roses tumble, also fragrant. In early autumn, apples and fat, juicy plums hang from another set of fruit trees, and the grape arbor is heavy with ripeness. In winter, the table has moved indoors, festooned with pine boughs and cones, with the paperwhite scent of snowy narcissus.

The table is set in the day for a long lunch leading well into the afternoon, until the day turns into evening, the sky slowly darkening and candles lighted in the encroaching night. Overhead light strings suspend with large clear bulbs that are not too bright. They mimic the color of the yellow flashes of lightning bugs. Somewhere an open fire blazes to cook on, to be warmed by. And all the while there is a certain kind of conviviality: There are stories, someone laughs, there is a good-natured but heated debate, a couple dances. This is the true nucleus of the farm, and particularly what we would call a biodynamic farm, the whole point of it really: the shared experience around the table that is defined by the cultivation of food, wine, friendship, ideas, and heart.

The idea of a farm-vineyard and all it might embrace came along with the thunderbolt: The name *la garagista* came to us immediately. Here, when you come to visit, or at tableside at our restaurant, I will tell you *la garagista* means "the woman who makes wine in her garage." And this will be true. But its meaning comes from several other quarters as well.

La garagista is a nod to the *garagiste* movement in France in the late 1970s and early 1980s, where a band of friends in Bordeaux decided to set out and prove to the world that it wasn't necessary to own a chateau and hectares of vineyard to produce world-class wine. They bought good fruit and made wine in their garages and sold their artisanal, small-batch bottles for steep prices. The idea of the garage or boutique winery was born, to be picked up later by savvy Californian winemakers. While I don't advocate that wine should be sold as just a luxury or an extravagance—I believe it should be part of our everyday ritual—I also believe that great wine can be made in garages and, in our case, supposedly unlikely regions.

La garagista also honors the dismantling of the old garage that once sat in front of our house and that has provided materials to make countless raised garden beds and an almost unending number of long farm tables that we can place end-to-end for big lunches or dinners, or where we pot up new seedlings, or rack small demijohns of wine. The space where the old garage was situated is now defined by a sunken, or secret, walled garden that cannot be seen from the road. The garden is enclosed by a curved stone wall with raised vegetable beds, fruit trees, grape vines, specimen plants, and one of those long tables where we now like to serve wine from our own fields in the milder months, a garden inspired from stumbling across old stone cellar holes replete with their own wild gardens cultivated by nature. This small, intimate garden where the old garage once stood is at the core of la garagista, and la garagista for us has become a sensibility, a way of living.

In that hot July, we planted those first one hundred grape vines. We chose the broadest slope of land. Our property is somewhat pie-shaped, pointing itself into a hop hornbeam wood. Almost all three sides of the wedge are bordered by both new- and old-growth trees: sugar maple, beech, ash, wild crab, birch, poplar. While this south- and east-facing piece is quite open to the sun all day, there are shadows cast later in the afternoon and in the evening at certain points in the season all around the edges of the land. In planting a vineyard, we've learned you need to be careful in watching where these kinds of shadows fall. The shade can affect vine growth and ripening, as well as the proliferation of fungal disease and certain detrimental insects—two of the biggest problems a winegrower can encounter.

When we chose the spot for the first one hundred vines, the section was at the top of the widest part of the land with trees nearby only on one side. The vines were settled into the breeziest part of the property, where they would receive full sun all day long. We knew that as we planted down our gentle hill, we could only go so far. As the property narrows, the ground does not drain as well, and the tops of those old, tall maples and beech begin to reach toward each other. Even though we knew so little that summer we

returned home with those young vines crowded in the back of the car, we knew we would be able to plant a total of about 2½ acres, or 1 hectare of vineyard here at our home farm.

The vines themselves were a mix of cold-climate survivors named Marquette, Frontenac, St. Croix, and La Crescent. These were the varieties that Lincoln Peak had on offer. We had tasted in the tasting room single-varietal wines or blends made with these grapes, and thought they were likely candidates. For the last fifteen or so years, these kinds of grapes have been grown up and down the Champlain Valley. We thought to plant twenty-five or so of each kind to see which thrived best in our soil. They all clearly thrived in the earth that was once the home to Lincoln Peak's strawberry fields. Chris Granstrom had been a strawberry farmer first before planting their land under vine. When Caleb and I lived in that area many years ago, we used to go pick strawberries there on a hot summer day, the small, sweet red fruit warm in our hands. Now, long and elegant bunches of grapes hang from leafy vines instead.

Over in the Chateauguay, there was no precedent for these kinds of grape vines. We sit at a higher altitude despite being slightly farther south. We had planted some St. Croix up a pergola the year before and knew those vines had done very well in the sheltered warmth near the house, but the open meadow was an unknown.

In a perfect world, we would have waited to plant the vines. We would have lightly broken the soil with tilling, seeded a cover crop, then tilled that back in the fall. Maybe seeded a winter grass of some kind like rye. We would have worked the soil again in the spring, sometime in April when the snow is finally gone from the ground. We would have planted at the end of May or early June after our last frost date. This would be the proper way to start a vineyard. Patience is the good winegrower's greatest virtue.

In this respect, winegrowing was perhaps not an auspicious vocation for me, as patience cannot be cataloged among my virtues. Impetuosity and stubbornness can. These qualities are good for taking leaps of faith, but not especially good for farmers, who must be steady and exhibit forbearance. Given that our farming endeavors have been trials by error—a textbook version of learning from one's mistakes—it is a good thing that another of my virtues is perseverance. Try. Try again.

So we planted at the wrong time, in improperly prepared soil, full of visions of harvesting our own wine.

In the early autumn, we held our first la garagista evening that spoke to this idea of the table at the center of the farm. We invited a friend who is an importer of Italian wines to bring a few cases of bottles to pour. We built a big fire on a broad stone at the end of our lower garden and grilled fresh bread and various vegetables like in an old-fashioned Italian *bruschetteria*. We invited about forty friends to taste wine and the spoils of that autumn's harvest out of the gardens, to spend an evening together outside in a farm-vineyard setting, even though the vineyard was just four rows large at that time. We were staking our claim.

After the larger group of friends left, long after the sun went down and a big harvest moon rose, we served a simple dinner to family and friends who'd helped coordinate the evening: a meal of soup and roast pork and an apple tart at a long table set with white linen in our barn. Candles were everywhere, and Caleb was convinced the whole thing would go up in flames. A friend passing through had brought fresh oysters from a trip to the coast to add to the dinner. We ate and drank enormously well; in this first realization of the farm-vineyard we were happy, and we knew, even though we had so much to learn, that integrating our lives into our landscape by growing food and wine was the next piece in the best part of our adventure yet.

BRUSCHETTA: RECIPES AND FIRE

Build a fire in a grill, in a fire pit, or on a stone. When the coals are hot, but not flaming, just as they should be when you are grilling meat, lay a grate over the fire for your sliced bread. The best kind of bread to use is a crusty, country loaf. If you are without a grate, you can use a pair of long-handled tongs to hold the bread piece by piece to toast.

When all the bread is toasted on both sides, even a little blackened on the edges, remove the slices from the fire. You can rub them with a clove of fresh garlic and just drizzle with olive oil, or add a variety of toppings: seasoned tomatoes, braised escarole, fatty prosciutto, or fresh radishes. Serve immediately.

Wind

When we moved to our place near the top of Mount Hunger, we didn't know about the wind, or that there was so much of it that our neighbor wanted to add windmills on the spine of his property. In choosing land for a vineyard, it is important to understand how the wind blows. This was something we learned about quickly and a natural element to our land that has worked in our favor. We could easily feel the airflow patterns from our first days living on this hill. *Air drainage* is the usual term for good air circulation within a vineyard, something that is very crucial to our northeastern grape growing due to our typically hot and humid summers. A lack of breeze can signify a wetter land, one that is more prone to fungal diseases, and grape vines are already very vulnerable to such complications.

AT THE SAME TIME, a vineyard must be guarded against excessive wind. Land that is too wide-open and flat can be subject to stiff northerly winds that bruise and batter the vine, breaking young stems or whipping off new green leaves. An open meadow with a hedge as a windbreak would be the ideal site. But the ideal is often elusive. If you are graciously presented with an open plain for a vineyard with so much sunlight, you take it, and plant a hedge of fir trees on the side where the wind most commonly comes to call. We've been lucky that our farm comprises an open meadow that is lined by hedges and trees.

In a more narrow tree-lined meadow, a vineyard should be situated closer to the center, out of the reach of shade. If the location is surrounded by forest with poor air circulation, the landscape can be cleared of a certain number of trees, or artificial methods like using a large blower to get the air moving around could be used if the season is exceedingly rainy or very still and humid. In these complicated scenarios, the winegrower who wants to husband the land as naturally as possible is confronted with difficult questions. To drastically change a landscape is to change the local ecology, which in turn has a ripple effect on the neighboring ecological communities: meadow, woodland, and wetland. And using an air blaster to circulate air in a still period of the summer burns up energy and natural resources. Which route has the least impact on the natural world around us? Does the winegrower decide that in the end this location is too problematic? Does the winegrower go in search of someone willing to lease her a better piece of land, or offer it in barter (wine down the road for use of the land) or some other creative solution? Or does the winegrower turn her creativity toward that air blaster? Could it be powered by solar panels? A small windmill? Used cooking oil from a local restaurant? Or would clearing or thinning some of the peripheral land actually benefit the local ecology, providing better sunlight and air circulation for all the immediate natural plants, since they too can be plagued by fungal diseases and predatory insects. The point here is that not all established vineyards were created on the best site. Sometimes, they were planted simply on the land that was available to the farmer. The questions asked and their answers must always reflect the responsible farmer's desire to create balance in the landscape.

Water

Of course, air drainage is not the only drainage to be considered. The drainage of soil plays a crucial part in the drama of the season. After a rain, if the ground dries up quickly, this is a good sign. If it remains boggy and sinks in as you walk along, this is not so encouraging. Some plants like to be planted in wetter areas; raspberries adore wet feet. Grape vines not so much. In land that retains water, adding drainage in the form of drainage ditches lined with stone and pipe or swales to help the water travel away from the vineyard, to be collected elsewhere, can be very helpful if not necessary.

WE DID NOT CHOOSE OUR LAND specifically for a vineyard; we chose it for its view and open sky. Once we decided to plant those vines, we needed to look at our land as a whole, and choose what would work best given what we had.

In our northeastern vineyard, lack of water is often not an issue; we are very rich in water resources. Our own land sits atop an aquifer, with springs under sections of the garden and vineyard. But given the shifts in climate that even we have noticed in the relatively short time we have set ourselves to farming, we have had hot and very dry seasons where we have experienced droughtlike conditions. Other years, we have had rain that seemingly would never end. It is critical to understand the topography of the land and the nature of the soil and where the water flows in order to manage the water in a vineyard appropriately. One of the best ways to understand how water moves

DIGGING FOR WATER

To understand how water drains in land, dig a few holes about a foot deep in the area that you want to check. Fill the holes with water and monitor how quickly it drains. Watch the holes after a rainstorm as well, to see how slowly or quickly the hole leaches water when the soil is saturated. This simple test will tell a lot about the water retention in the soil. I recommend digging these holes in several places in a plot set aside for vines or any other plants.

through land is to dowse. Most people think you need to hire someone adept in dowsing or with a special sensibility. Some people think there is a certain hocus-pocus around the idea of dowsing, that only modern pagans believe in such things. But we all have the ability to dowse. All that is needed are very simple tools and a little guidance.

HOW TO DOWSE

GENERAL DOWSING IS RELATIVELY SIMPLE. For exact and thorough information, work with a professional dowser, but you can at least get a good idea of where the water flows on a piece of land by doing it yourself.

Take two thin copper wires, the same size as what you would find in 12- to 14- gauge coated electrical wire. Cut two pieces, each about a foot in length, with wire cutters. Make two 90-degree bends in each wire.

To dowse, hold the wires loosely but steadily in your hands, your elbows bent and your forearms parallel to the ground and facing straight ahead. Walk across the land slowly, especially near areas that you suspect drain less well. The wires will begin to move and will cross each other when you cross water. Mark the area for reference.

Soil

The choosing of our vineyard site within the larger context of our farm was relatively easy based on light, air, and water flow. However, understanding the soil is critical. Lovers of wine have always heard that the soil in a vineyard is directly related to that word *terroir* and how significant it is to the vine. Lovers of wine have also heard that vines love poor soil and thrive in poor conditions. By nature, they must struggle in order to make wine of character. You wouldn't want to grow vines in soil appropriate for the vegetable garden, soft earth full of organic matter and humus.

BUT WHAT EXACTLY does *poor soil* mean in this instance? Why wouldn't any plant want to grow in a well-nourished dirt? Poor soil, in this case, is a bit of a misnomer, and can be misleading. Grapes love *well-drained* soil. Rocky, sandy, porous. They do not like excessive water. This notion of poor soil seems to have been proliferated by photographs of famous vineyards in France where vines are planted in the remnants of prehistoric riverbeds, the ground covered in tennis-ball-sized stones. The whole vineyard looks like a prettily arranged and very wide drainage ditch.

Of course, the vines love this kind of drainage, and they love the passive solar heat collected by the stones, but the image doesn't take into account what might be happening beneath those stones.

Some varietals prefer sandy soil compositions, while others prefer land with a bit more clay content. Wine will develop different characteristics when grown in different soils. I have two wines at the osteria made from the same grape, Dolcetto, but grown in different areas and soil types. The Dolcetto grown outside of the town of Barolo, parcels of land defined more by clay and calcareous components, is very different from the Dolcetto grown in the pocket of the Langhe. The soils there are more limestone-rich and sandy. The Dolcetto grown in the clay is brawny, inky, very masculine, and the Dolcetto grown in the sandier soil is all about bright, fresh fruit, and more feminine.

Each grape varietal is different and likes slightly different circumstances—at the same time, whatever soil it grows in will affect the wine that's made from it. All vines, though, prefer having some stony material as part of the soil makeup, as they need that good drainage.

One of the ideas about good drainage is that the surface water runs off and down through the top layers, so the vines must dig much deeper to get to a water source. Just because they like good drainage doesn't mean they don't want or need water. Vines cannot be grown naturally in a desert. If the water source sits near the surface of the soil, that's where the vine's roots will stay. Nature is no dummy, and the vine will take the easiest and most efficient course for its survival. But leaving the roots near the surface of the earth makes for a lazy, less hardy plant. One that's more easily uprooted, more easily affected by disease and pests. It's like a vine with a bull's-eye on it to the natural world of predation. These vines also tend to make uninteresting wine. They make flabby, indifferent wines with fruit and juice more susceptible to problems in the cellar. But if the vine must work its way down into different strata of rock and soil, down deeper into the ground, it gains strength and feeds on more minerals. This is why it's always said a vine must struggle—its longest taproot and accompanying roots must work to feed the plant, and in feeding from a more complex plate, the wine itself becomes more complex. Never has the maxim "you are what eat" been truer.

But while those roots dig deep, there are still smaller, hairier roots that remain closer to the soil surface in younger plants, providing the plant with a different and just-as-essential kind of nourishment. And while the wine-grower encourages the plants to dig deep into that more complex soil, the farmer can't forget that top layer at the surface. That topsoil must remain healthy and full of life to feed the young vine as well as have nutrients that are carried deeper and throughout the soil by all kinds of little helpers that live in the topsoil layer.

No one ever talks about young vines. Wine talk is usually centered around vines that are ten years old and older. Viticulture discussions seem to always revolve around established vineyards and the work that's done after those first three crucial years of a vineyard's life. There may be a passing reference to how you might differ your program for the start-up vineyard, but there are few books that illustrate how best to care for the newly planted vines.

Fortunately, a few good resources do exist. *The Grape Grower: A Guide to Organic Viticulture* by the late Lon Rombaugh is a classic text on how to put together your first vineyard. Yet this kind of book can be confusing to the wine drinker turned winegrower, or the wine drinker who has an interest in how it all works but who's been taught the idea that poor soil is supposed to make such great wine.

The soil is of key importance in organic viticulture. The dirt drives the plant, feeds the plant, defines that plant. The climate, the grape varietal, the geography, and the grower are all part of this equation too. But the health and geology of the vineyard is where it all begins.

In a perfect vineyard world, the winegrower takes two years to prepare the site. Preparation always begins with an analysis and reading of the site. The local county agriculture extension can help prepare a soil sample based on their requirements. Generally, dig a hole about a foot down and collect a handful of soil. Do that in about ten or so spots throughout the vineyard site, just like checking for drainage. Then pack that sample off to the extension for soil analysis. The results will tell what the mineral composition of the soil is, its pH or degree of acidity, and whether the soil is undernourished or over-nourished, plus they will provide basic information on how to correct nutrient deficiencies or overabundances. The idea is to create a balanced soil for the growth of the fruit. Typically, state- or university-run agricultural offices tend to offer recommendations that may be tailored for a more conventional growing program rather than for organic methods, but that office can often also make good suggestions of where to go for organic information.

This is exactly what we did in the autumn after we planted those first one hundred vines. We didn't know much about our soil other than the obvious: We had loam and we had clay. Some sections of the meadow were also sandier than others. When the tests came back, we saw that our soil was deficient in everything: low in magnesium, phosphorus, and nitrogen. Plus, the pH was highly acidic. Perfect for wild alpine berries and ferns, but not so kind to grape vines. Indeed, work needed to be done.

Once we had planted these first vines and got the report back from the agriculture extension, we knew we had stepped over a threshold. We had these vines in our care now, and the soil was offering very poor nutrition. There was no turning back. We had to learn what to do.

Soil analysis is not the only way to gauge the soil. You can also assess the state of a soil by what grows there. What grows in a plot of land can tell a powerful narrative of the soil, one that is deeper and even more helpful than the soil test performed in the lab. If dandelions grow on the land in the spring, this is a telltale sign that the soil is compacted. Dandelions only grow in compressed soil; they are nature's plow. Those extravagantly long and tough taproots that refuse to leave the ground when you try to pull them up are there for a reason. They push through and break up the soil, and when the plant dies, the root withers, leaving behind a cavity, or *macropore*, that allows air, water, and microorganisms to pass through it. The wild alpine strawberries that covered my own vineyard floor told me that my soil was highly acidic, and as beautiful and precious as these small plants are, the sheer number of them would have to diminish for the vineyard to thrive.

Reading Botany

A few winters ago and another 350 grape vines in the ground, I applied for a grant for education and business development that supports farm women in Vermont. An open grant, this award from the Vermont Farm Women's Fund allowed me to go study biodynamic viticulture in Burgundy.

I had been following and learning about biodynamics in the vineyard through my work as a sommelier. I wanted to educate myself more in relation to my work in the vineyard.

THAT SPRING WE PACKED OFF to Puligny-Montrachet where I would meet Bruno Weiller, an Alsatian man who had a history in growing wine. Now he specializes in biodynamic farming and consulting for vineyards, his focus the relationship between the soil and the vine, and I would learn from him the essentials of biodynamics in the vineyard. In reality, despite Rudolf Steiner's organization of biodynamic thought, there really is no codified system, no prescriptive plan that everyone can follow. Each winegrower must respond to the land, the plants, the weather. You must make decisions based on what you see, hear, smell, taste, and feel. This is a mantra I repeat constantly to myself. This is farming at its core and with intuition. It is farming in the way we farmed thousands of years ago, when the first hunters became gatherers, and when we could only rely on our senses and on trial with error for our schooling.

A SQUARE IN PULIGNY

Puligny is lovely in the spring. It is a well-laid-out provincial town with high stone walls and a greensward more typical of New England than rural France. That spring the sun was warm, the vines just starting to leaf out. The vineyards looked tended, and there was always someone out quietly working among their gnarled, yet graceful shapes.

My course was held at *Ecole du Vin et des Terroir,* a program started by the rather infamous wine producer Anne-Claude LeFlaive, her notoriety coming from her inheritance of failing vineyards that were at the point of dying, old vines that had been treated conventionally for a number of years and were so sickly that she was advised to pull them out and plant anew.

Anne chose to try to save the vineyards instead. She threw herself into the study of biodynamics and how it might help her turn these parcels around. Within three years of intensive biodynamic work, the vineyards sprang back to life.

Like any convert, she felt she had a mission to help others bring that kind of positive energy into their vineyards, and so the *Ecole* was born, offering workshops and courses to seasoned winegrowers and neophytes alike.

When I applied for the grant, I understood that courses could be chosen in English or French. I had studied French for many years, but my time living and working in Italy had shifted my language strengths. Italian had superimposed itself over all that *passé composée.* I chose my course as an

English-speaking one. The course was slated for April. In January, I found out that they were very sorry, but the course would be taught in French. However, the course leader would be able to speak English if I needed additional explanation.

Not to be daunted by a mere language barrier, I spent three months reacquainting myself with my French studies and building a French vocabulary in viticulture, and by the time we boarded the plane I felt tolerably proficient, and that I had enough language to get by.

We stayed in a little hotel on a square in the prosperous wine town of Beaune in Burgundy. The linden trees were in flower, and we ate Burgundian classics like *ouefs en meurette* and ham in aspic in the hotel restaurant. On the first day of my course, Caleb dropped me off at the gate of the program; dressed in jeans and Wellingtons, as a large part of study would be in the field, I somehow felt about five years old again, being left at school for the first time. I was one of the first to arrive for the class.

Everyone was very kind to the aspiring winegrower from America. Bruno Weiller, the leader, especially. He did, in fact, speak English, and so did the program administrator who would be sitting in on the class. They assured me I could ask questions in English if anything went above my language skills. I felt completely relieved and at ease until we all sat down in the classroom in a semicircle with Bruno in the center.

There were twelve of us. Most everyone in the group was a longtime winegrower converting from organic to biodynamic. One gentleman was from Switzerland. Another woman was a *sommelier-consultante*. She bought small quantities of wines from small producers all over the country. She would hold dinner parties in her home for ten to twenty people every time she got back from a wine-buying trip, showing the wines she had procured. Her clients would dine and taste and buy. She specialized in biodynamic wines and wanted to learn more about the process in the vineyard. Another of the women was the winegrower for a smaller organic vineyard owned by a larger industrial producer. Already at that time, biodynamics in winegrowing was becoming de rigueur and part of the marketing strategy for many larger wine producers.

As an icebreaker and a gauge, Bruno asked everyone to say a little bit about who they were, their background, and why they were here. My temporary relief ebbed a little as I realized that I was going to have to say something in front of this group, in French, that sounded reasonably intelligent. My relief ebbed a lot more when Bruno looked to me and winked and said that I could start.

Mon mari et moi, nous avons une petite vigne nouveau entre les montagnes en Vermont, en Etats Unis. . . Je suis une vigneronne. "My husband and I have a small vineyard in the mountains of Vermont in the United States. . . . I am a winegrower."

These were early days to claim such a thing. But as if overnight, and in the utterance of one sentence, it had become almost true. The simple purchase and planting of the original vines on our parcel made me so. The planting of an additional three hundred plants the following year with another three hundred plants to be put in the ground upon my return from France would keep me out of trouble, or landed more fully in it, depending on how you looked at it.

Somehow, I managed over those several days. All my senses sponged the atmosphere, the conversation, the information. I muddled along in speaking French, though my comprehension was, as it always had been, stronger. Bruno was there to answer any question, as were a few others who also spoke English. The sommelier-consultante had actually taught English in a former

FLORA IN BURGUNDY

career. By the end of the course, I was exhausted by all I had absorbed. When we returned from the field on the final day, we sat in that semicircle again and Bruno asked us to talk about what we had gained from these last few days and what we thought of the format, given that it was the first time this course had been taught. This time I got to go last, but by that point my brain was so dog-tired I could barely put together a whole sentence in English, let alone French.

The course had centered on botany and soil. We went out into the *premier cru* and *grand cru* vineyards owned by the Domaine LeFlaive surrounding the village. We looked at, smelled, and even tasted soil. We compared it with neighboring soil that had been treated chemically. We talked about structure, humidity, health. We began to understand something about soil that was alive and soil that was without life. We collected plants in the vineyards as well as in the fields fringing the parcels. We spent hours dissecting the natures of dandelion, horsetail, nettle, and other field flowers. We learned how these plants interact with their environment, what their growth habits tell us about the soil in which they grow, and what their intrinsic characteristics might offer in terms of plant medicine to be used at certain times and under certain conditions.

Certainly Rudolf Steiner's work with biodynamics did systematize seven plants to be used during the growing season: dandelion, valerian, yarrow, stinging nettle, horsetail, chamomile, and oakbark. And we studied those seven plants rigorously throughout the course. But Bruno also had us collect other flora from the fields to see if we could dissect their intrinsic natures, and, from what we learned, to see if we could extrapolate a way in which we might use them in plant tea preparations to help provide a boost in the vineyard.

This was the most challenging work of all. To understand the absolute nature of a plant and how it fits into an intimate ecology involves a lifetime of study. This was Bruno's point: It takes a lifetime to understand your landscape, and you must take the principal ideas of biodynamics and apply them to your own environment. What might work for the winegrower in Switzerland might not work for me in Vermont. We may be able to share in some of the applications, like those structured seven plant tea preparations, but each environment will be unique and ever changing. This is biodynamics on the soil level.

When I returned home from that visit to Burgundy, I started to make a catalog of the plants growing in our vineyard. After two years of light tilling between our already planted rows—and, as we worked the land, some judicious applications of composted manure—the flora had begun to shift. The sensitive fern, another indicator that the soil was acidic, had disappeared, and the poplar striding from the hedgerow no longer threatened. In their place, we saw the proliferation of native red, purple, and white clover, which suggested that the levels of magnesium and phosphorus had improved. The green compost of the clover, as it is cut and worked back into the soil, offers a life-giving dose of nitrogen. A proliferation of English plantain has been further encouraging. Plantain has an affinity for vines. It is one of the first plants to arrive after tilling. Plantain protects disturbed land from erosion and attracts birds into the vineyard for food. While we do worry about birds eating fruit at the end of the season, birds eat insects, too, and are part of a beneficial cycle in the vineyard.

I ALSO BEGAN TO LEARN what plants grew when. After the soil had been turned, we would often see an explosion of pigweed, or lamb's quarters, which is a close relative of spinach that has been used in salads in Europe for thousands of years. It is a plant very rich in vitamins A, C, and some B's, and contains calcium, iron, and phosphorus. While we, too, pick the lamb's quarters from the vineyard for salads at the osteria or on the farm table, we also know that hoeing those plants back into the soil will help the soil to balance its calcium, iron, and phosphorus—all nutrients our soil was very low in when we began reclaiming the meadow.

The Ground Up: Cover Crops

Once we were connected to the soil for our vineyard through a soil analysis and some study of what is currently growing here and what that means, the soil preparation for new blocks began. Many winegrowers plow and till; others do not. Some growers have parcels of land that benefit greatly from gently breaking up the soil to encourage light and air in that top layer. Other producers feel that such a breaking of the soil would disrupt and destroy the finely tuned microbial activity. Both growers are right; the choice depends on the circumstances of your land and the climate. For us here in the Northeast, to work organically in locations with a percentage of clay, plowing then tilling the land is the first order of business. Planting a cover crop that gets tilled back into the ground helps feed soil that hasn't been worked in a long time or soil that has been chemically treated. We like buckwheat as a summer cover crop, as it makes a very fragrant white flower that the bees love, and if you are patient and have the time and inclination, you can collect the seeds to grind for flour or as groats. But there is a litany of other grains, legumes, and wildflowers that can be planted as well.

RECIPE FOR
WILD VINEYARD SALAD

Learn to identify a few wild greens that grow in meadows in your area. Dandelion is a good spring green to start with, as it grows in most places; the tight, pale green bud surrounded by the lion's mane of leafy greens is also fairly easy to single out. In our vineyard, we look for and pick dandelion, wild chicory, wild-seeded arugula, purslane, and pigweed. If any of the plants have gone to flower, we pick the blossoms, too, to toss into the salad. Pick the leaves by cutting them at the base of the plant, and rinse them well in cold water. Spin dry. Toss with a classic vinaigrette made with Dijon mustard or a dash of balsamic vinegar, the sweetness playing off the bitterness of the leaves.

CLASSIC VINAIGRETTE

1 clove garlic

1 part red wine or white wine vinegar

1 teaspoon Dijon mustard

3 parts mild organic olive oil
  or organic sunflower oil*

Salt and pepper to taste

Rub the walls and bottom of a wooden or pottery mixing bowl with the clove of garlic. Add 1 part red or white vinegar or a blend of both. Add a small teaspoon of the Dijon mustard. Add 3 parts of the oil, and salt and pepper to taste. Whisk with a fork or a whisk. Taste. Add an extra dash of any of the ingredients you prefer.

A NOTE ON THE OIL: We learned to make vinaigrette in Italy as well as from the tableside waiter we had one night at the restaurant Le Continentale in Quebec City. In choosing the oil for your dressing, be aware of the balance of flavors of the oil and the vinegar. In Tuscany, where we used to live, they predominantly use sunflower oil for vinaigrette because it does not overpower the flavors of the vinegar, and sunflowers for oil grow in abundance in the countryside. Tuscan olive oil tends to be very strong in flavor and rather spicy, which would overwhelm the dressing. Other areas in Italy and France also use sunflower oil, and those areas that grow milder-tasting olives will use the olive oil as a component in the dressing.

COVER CROPS ARE ESSENTIAL in good organic or biodynamic vineyard management. Planting a cover crop, or various cover crops, keeps the vineyard from turning into a monoculture of one plant, the vine. Typical cover crops belong to the Fabaceae, the legume plant family: fava beans, sweet peas, bell beans, or clover, among others; all are beneficial for vines because of their ability to forge a symbiotic relationship with nitrogen-fixing bacteria in soil. When these plants are tilled back into the soil, they decompose and feed nitrogen in a usable form to the grape vines, improving fertility and invigorating the plants.

Cover crops can also regulate vine growth by de-vigorating. Plants that are nonlegumes can provide root competition for nutrients and water for varietal vines that tend to be overzealous in their growth habit.

In the case of leguminous plants, besides increasing the nitrogen in the soil, decomposed cover crops can increase the soil's ability to exchange and hold nutrients. Nutrients from plants are often chelated—that is, they form complex organic compounds that are more readily exchanged between plants than the inorganic synthetics that make up conventional chemical fertilizers. Many organic growers also apply compost, which additionally increases the fertility and the ability of the soil to redirect the nutrients properly to the vines.

WILDFLOWERS AS A COVER CROP

BUCKWHEAT COVER CROP

Soil organisms that use decomposing cover crops as a food source create waxes and other sticky substances that hold the fine particles of the soil together, forming aggregates and improving the soil tilth, or structure. As the organic matter increases in the soil, so does its ability to hold water. Fine root systems also hold the earth in place, keeping the soil stable and preventing erosion. Nitrogen formed by legumes is less mobile and less soluble than chemical nitrogen fertilizers. When cover crops assimilate free nutrients in the soil, they stabilize them during periods of high rainfall. During dry periods, the cover crops also reduce dust and improve air quality, as well as the possibility of mite infestations, which can intensify under dry, dusty conditions.

Another good cover crop to consider is that of grains. In the autumn here in the Northeast, farmers often plant cereal rye that will grow during cold and inhospitable weather in the fall. These plants contain carbon and aid the vineyard soil by increasing its tilth as well as its porosity and aeration. Often a combination or a cycle of cereals and legumes is used to benefit the vineyard in both ways, and depending on the site, varietals of vines, weather patterns, and vegetative quality of the vines, annual or perennial types can be chosen. Allowing perennial plants and grasses to flower before tilling or cutting increases the carbohydrates in their root systems, which improves their persistence and competiveness with unwanted plants.

Cover crops also draw in excess moisture. While it is important to look for well-draining soils, there can be sections of a vineyard that drain better than others. Our own vineyard is wetter on the north side and much drier on the southwest side. While some cover crops can help a dry vineyard retain water, planting different crops can help soak up excessive rain and unwanted groundwater. Oats respond well in more clayey, water-retentive soils.

One of the objectives of organic winegrowing is that plants will have larger root systems that dig deeper into the earth. Because soil nutrient concentrations from cover crops are lower but cover a larger area than synthetics from conventional fertilizers, root systems are encouraged to extend and gather nutrients from farther afield.

Organic matter in the soil, or belowground, is a food source for both micro- and macroorganisms. Many of these organisms assist in recycling cover crops into the soil, improving the soil's physical attributes at the same time. The rise of earthworm populations is a good sign of soil health and improving conditions. Much research has been done on the incorporation of organic matter into the soil from cover crops, and it shows that biological organisms can reduce damage from root pathogens by inhibiting their growth and development.

Aboveground, cover crops produce flowers, and flowers in turn attract various beneficial insects that attack invasive species that can do damage to the fruit or the vine. Mites, mealybugs, galls, and midges can be controlled very effectively and naturally this way. When working organically, it is crucial to understand that insects and fungi, all the things that plague the grape vine, cannot be eradicated, nor should they be. They live in the soil and are part of an intricate chain of ecology that creates balance. The goal of any grower should be to control these pests by keeping them happily in the ground, or, if they have already begun to beset the plants, to control the amount of damage that can be done. Even the intensive chemical insecticides and fungicides will not remove the threat. They may offer a significant amount of control in initial applications, but insects and fungi are clever organisms that respond to the chemical structures of these pesticides; the insects and fungi mutate, change, and shift their agendas, rendering the pesticide less and less effective over time. In contrast, the cover crop works naturally, providing a habitat and a food source for beneficial predator species. Scientifically the relationship between predator and prey (pest) within the environment of the cover crop is not entirely clear, but reduced problems from insects are apparent when cover crops are planted instead of using conventional insecticide applications.

Cover crops also help control invasive plants not wanted in the vineyard by edging them out over time. While grasses can be effective in the first three years of cover cropping in various locations, too much grass is not desirable in the vineyard. It can eventually influence the taste of the fruit and the taste of the wine, so planting other things can help restore balance to the flora.

Cover cropping has many advantages for the soil and the plants, but it also has an economic edge. Cover crops create a natural source of nitrogen for the soil, so the need for outside fertilizer inputs is reduced. Focusing on cover crops as a way to manage the health of the soil helps a farmer move away from herbicides and pesticides as well. And because the soil becomes more vibrant, the quality of the yields for a specific cash crop becomes that much greater. Since cover crops also help the soil retain moisture and prevent erosion, larger-scale farms tend to have a crop-rotation plan that incorporates seasonal and yearly sowing of crops in relation to the farm's cash crops. But in our small-farm situation, we have found that in many cases, we only need to seed every other year, as the flowers of the plants will reseed themselves. You can choose from scores of various cover crops that might suit the landscape best, but it is also possible to choose a volunteer cover crop, one that naturally grows on the site. In Vermont clovers are abundant, and they provide a very good choice. Once land is broken by tilling, clovers are one of the first plants to start making their grand entrance.

Every third year, we intentionally don't reseed the cover crop so that we can see what mixture of plants will grow in the soil. This allows us to see how effective the cover crops have been, as certain plants will only tolerate certain

POPPIES AS A COVER CROP

conditions. This year off from cover cropping helps us by showing us what the soil needs, and helps us choose the kind of crop we'll plant in the fourth year of a rotation.

There is so much to recommend cover crops, but it is important not to forget that the animal world can also have a bolstering effect on any piece of land, especially on areas that have been out of use. Putting animals on an open piece of land—sheep or cows to graze for a season, or pigs to wallow and till with their snouts—brings natural compost and energy to places that have lain fallow for years or have been allowed to become overgrown. Having animals present one to two years before planting, to properly prepare the soil, is the preferred method for starting a vineyard off in good fettle.

Over the years, we have been planting our vineyard in blocks in succession, which has given us time to ready each section. Cover crops, like our favored buckwheat or sunflowers, make beautiful cut flowers in the osteria and in the farmhouse, as well as providing grain and seeds for cooking or feeding wildlife. More recently, we have begun to convert new plots every year or two before planting by using them as a market-style garden, planting particular vegetables for the osteria that work well in the sandy or clayey soils on our land and may need room to stretch out. Chicories, radishes, ground cherries, mustard greens, squash, cucumbers, and carrots have all been welcome, and prepare the ground very capably.

Historically, in Italy, farmers used to plant such crops between every other vine row. In the Piemonte, they would plant corn for polenta and wheat for flour. As our vineyard is much like a garden and has in fact become an extension of our garden, either for its wild edibles harvested for dishes, or for the market-style gardens surrounding the plots, we have begun to use more readily edible crops as our covers. Planting daikon and carrots in every other row keeps the soil from getting too compacted, and flowers like nasturtium, poppies, and cosmos all make for delicious cut flowers and edibles for the vases and plates at the osteria, or at our farm table.

Vines, Varietals

In our hillside meadow, 1,600 feet above sea level, we must grow alpine wines. Both our landscape and the currently prevailing climate dictate that we must grow varietals that can manage the cold, snowy winters and hot, humid summers. In these northeasterly climes, those crosses made between indigenous vine stock and *Vitis vinifera* vines from Europe are perfectly suited for our conditions. After the first very experimental years of vines growing on our land, we chose to focus on four varietals: the Marquette and La Crescent for cold-hardiness, and for experimentation, the traditional cool-climate Riesling and Blaufränkisch, "Blue French," a red typically grown in Austria. There is not much proven *vinifera* here in this little slice of the world, but we decided that we should try, even if it meant that we didn't get a full harvest of fruit every year.

WE HAVE SEVERAL ROWS of unusual and intriguing blending grapes as well, all cold-hardy: Frontenac, Frontenac Gris, Frontenac Blanc, and St. Croix. They have lighter, fruitier personalities that we now pick for a home farm field blend, lending a winsomeness to the wine.

At the edge of the vineyard, near where we grow potatoes, we have another experimental planting of Melon de Bourgogne, or Muscadet vines. Known for its frost resistance, I fell to the temptation of the name, or the desire of my palate to make my own dry, mineral-laden, saline, Loire Valley–

style white. Chances are it will go the way of the field blend, when the vintage is kind enough to give us a harvest, adding its own personality to a complex mélange of flavors and textures.

The two primary varietals we grow are long on character. Marquette is generally a dark wine with intense pigmentation and flavor notes of mint, black pepper, black cherry, savory beef, and the dusty sweetness of black-berries. It can speak of earth or the soft, fragrant petals of a rose, and sometimes the two together. It is striking. It is sanguine in nature. The first time I tasted La

PETITE PEARL VINES

Crescent, made by another local producer, I imagined making a wine like a finely pearled Moscato from the Piemonte in Italy. Off-dry, the sweet tea and floral character would skip right down the tongue. On further acquaintance, La Crescent has a deeper voice, and the wine that comes from the plots we farm here and in the Champlain Valley beckons for the juice to be in contact with the grape skins and oxygen during the initial fermentation. The result is a very dry, orange-styled wine with many layers of flavor, an unfolding story of crushed orange flowers, white pepper, olives, orange rind, bergamot tea, and the waxiness of a church candle. The development of tannin, aroma, and texture from the fermentation style shows itself as mysterious, yet there are enough top notes that keep the wine lighthearted and delicious.

The other cold-hardy varietals each have their own voice to offer. Frontenac Noir on its own can have a lot of woodland berry fruit with a juicy quality, but is usually made with residual sugar; the Noir becomes leaner and more mercurial when fermented completely dry. Frontenac Gris with its pink and gray skins calls to mind a cross between young Sauvignon Blanc and Pinot Gris in its citrus acidity that loves to be paired with food. Frontenac Blanc, meanwhile, adds notes of lemon and exhibits clarity, like a bell clanging. The rather obscure Brianna that we have inherited from one of the vineyards in the Champlain Valley is a little tropical and fleshy, and is saved from being blowsy by its dryness and slightly bitter acidity.

It seems like almost every day there is a new cold-hardy cross available. They are marked by strange combinations of letters and numbers before they can even be named, like MN1094 or ES 4-7-26. So much energy is being put into the development of cooler wine regions since the success of places like Quebec, the Finger Lakes in New York, and northwestern Canada. This year's little pretty on offer is Petite Pearl, a light red touted as a good blending companion to Marquette because it is low in acidity and bright in fruit. The nursery where I have bought vines is eager for me to plant some, but much like how I approach my work as a sommelier at the osteria, I prefer to taste a wine before committing to it in the glass or in the ground. At the osteria, I need to understand how a wine works within my wine list and how it will respond to the ingredients and food on our menu—in other words, Caleb's cooking. Just because a wine tastes good to me doesn't mean it is right for our list.

My feeling is the same in the vineyard and for what I choose to plant. I'd like to at least taste the Petite Pearl fruit, if not a wine made from its grapes. I need to know if its essential character will fit in this environment and if it will integrate with the wines we make here.

We plant wines that we love, that intrigue us, that call to us in some way. But these are also grape vines that are appropriate to our terroir. It would be a mistake to plant Californian varietals in northern Vermont, and it is a double gaffe to choose the correct varietals that will survive well in a landscape, but then to try to bend and shape the wine into some preconceived notion of what a similar wine from elsewhere might taste like. What sense would there be in trying to make California-style wines in the Northeast? It is key to look to winemaking regions with similar terroir. So, we turn ourselves to Austria, Germany, New York State, Quebec, eastern Europe, and northern and alpine France and Italy.

Planting, Preparation

Planting in our northern clime comes later in the season. In more temperate locations planting can take place in the fall or early spring when plants are still dormant. Here, we plant after the last frost warnings so that the bitter cold does not blacken the edges of newly unfurled leaves or hopeful buds. Yet the vagaries of our weather patterns over the last several seasons have also served to confound us. Very warm Marches and early Aprils get us out into the vineyard and gardens earlier, but these warm days entice wickedly. Just when you think it might be safe to flout the conventional wisdom, the weather turns cold and very rainy, foiling the pollination of the already blossoming fruit trees. They have not been so lucky for a couple of years: the stormy, cool weather driving the bees back to their hives, and the cold and wet conditions destroying what pollen was available. One year, we had no tree fruit at all.

SO WE WAIT. We learn the lesson from the fruit trees and curb our desire to get young vines in the ground. When farming biodynamically, you look to certain times that are most propitious for planting, cultivating, harvesting. You follow a calendar that maps out days good for certain types of plants. For planting, we look for days that are particularly auspicious for perennials that produce fruit. We look for the stars clustered in Leo to be in the sky. Historically, this is a time most propitious for the growth of grape vines.

Nonpractitioners of biodynamics often misunderstand the planting calendars, referring to them as planting by astrology. The calendars are not attached to astrology but rather astronomy, dictated by when the moon and sun and the other planets are in certain cycles and certain distances from the Earth. The same kind of cycles as the tides of the sea that connect to the rising and setting of the moon.

The last frost date is usually around end of May. We prepare the soil by tilling those market garden beds earmarked for planting earlier in the spring, then hit them one more time closer to planting. This creates the best planting situation. In the past, we've used augurs to dig holes, we've hand-shoveled or used a posthole digger, and while we've always tilled and broken up the ground before in those instances, the soil has not been as amenable to planting as that prepared in the market-style garden beds. These blocks are made of soft, aerated soil of good tilth.

In the past we have ordered young vines from nurseries to get us started. Now we are at a point in the growth of our vines where we can take cuttings from our own stock, or *selection massale*. Many winegrowers believe that selection massale helps maintain and retain the identity of a particular vineyard site by reproducing vines from the best vines that already exist there. If clones from outside sources continue to plant and replant a site, then the vines are no longer the vines that have adapted well and symbiotically with a given terroir.

When we prune, we choose 12- to 18-inch straight prunings from the first 1 to 2 feet of new wood for the cuttings. This wood will have buds or nodes closer together. This is beneficial because most new roots develop from nodes. Also, new wood that is closer to the old wood of the vine will have more nutrients stored in its system, making for a healthier cutting. Typically, we look for cuttings with a diameter of a pencil, up to ¾ inch thick, though American hybrids can still do well if they are smaller. We make the cut on the bottom flat so they are easier to bundle, and the cut on the top at a slant so that we know which is which. The bottom cut should be ¼ inch or less below the bud, and the top cut should be about 1 inch above the bud. Given where we live and when we prune, we typically don't store the cuttings for very long. For any storage we keep them in plastic bags with a damp cloth to keep the atmospheric environment moist. They should be kept in a space cooled to between 32 and 36 degrees Fahrenheit. When we are ready to root, we soak the cuttings in water overnight to let them suck up as much water as they want, and "reactivate." We dip them in a powdered or liquid root inoculant like a mycorrhiza, a beneficial fungus used in organic gardening, and set them out in a holding bed made of potting soil. It takes about two to four weeks for them to root. We keep the soil moist, but try not to overwater during this stage, as the water will slow down the rooting as it cools the root zone of the cutting. Once they have rooted, they go out into the vineyard.

When we go to plant in the vineyard, we dip the roots in a mycorrhizal inoculant again, then use a straight piece of iron rebar to dig the hole, and in the vine goes, the roots below the ground, the earth hilled and packed up a little around the vine. A good watering follows if there is no rain in the forecast. Hilling and packing the dirt around the vine ensures that the soil doesn't sink below land grade after watering, making a depression or well that could eventually expose the root structure of the plant.

Mycorrhiza is a very important part of the planting process. The word comes from the Greek *mykós* for "fungus" and *riza* for "roots." A mycorrhiza is in fact a beneficial fungus that has a symbiotic relationship with the roots of vascular plants. The roots and the fungus form a mutualistic alliance. Mutualism means the way two organisms of different species exist in a bond in which each individual benefits. Pollination is a classic example of mutualism.

Mycorrizhae form this mutualistic relationship with the roots of most plant species. Vines particularly thrive in this dynamic. The mutualistic association between the vine and the fungus provides the fungus with a relatively constant and direct access to carbohydrates like glucose and sucrose that are generated in the leaves of the plant and travel down the vascular system to the root tissue where they feed the fungus. In return, the plant receives a higher absorptive capacity for water and mineral nutrients due to the comparatively large surface area of the mycelium. (The mycelium is the vegetative part of a fungus, a mass of branching, threadlike filaments.) This extensive system greatly improves the plant's mineral absorption, a primary goal in winegrowing, connecting the vines even more intimately with their terroir.

Mycorrhizae make a very beneficial root inoculant for planting new vines, and their use is standard practice in organic and biodynamic programs. Many conventional vineyards plant using a synthetic fertilizer to pump the new vine with food and encourage it to grow quickly in that first season.

RIESLING ROOTS

Feeding your vines with things like compost, which delivers natural nutrients more slowly over time, is a very good thing early on, but a chemical application of fertilizer only becomes a disservice to the plant after the first season. Using chemical fertilizers babies the vines, providing easy food on the surface of the soil. The roots have no need to delve more deeply into their habitat, so they grow outward along the surface. Because the roots are shallow, there is very little stability, and they are not given the chance to interact with the mineral nutrients farther down the strata.

Organic mycorrhizae can be purchased from a garden nursery that offers organic products. There are companies that have made it their business to grow these fungi for improving soil structure and aiding plants. In biodynamics, a similar mycorrhiza is used, but it is made on the farm.

Almost everyone who knows just a little about biodynamics knows of the practice of burying a cow horn in the ground for the winter. It is one of the preparations that makes many consider biodynamics some kind of voodoo. The reality is that these very old preparations are how things were made before we had labs or research nurseries or university horticulture programs. The internment of a cow horn or a stag's bladder may seem mystical and paganlike to the modern sensibility, but much of the work that is being done now in laboratories would seem the same to the farmer at the turn of the last century.

One of the primary preparations in biodynamics is cow manure stuffed into the horn of a cow that has had at least one calf and lactated. You can tell how many calves a cow has had by the rings on her horn, just like telling

ROOTS DIPPED BEFORE PLANTING IN A MYCORRHIZAL SOLUTION

the age of a tree by the rings in the trunk. Horns are important on cows as they are part of a complex system that trigger the creation of certain kinds of enzymes in the milk. The cow horn from a more mature cow is then buried in the fall, around the equinox, and left over the winter. In the late spring, sometime in the latter half of May, it is dug up. By that time the manure in the horn has turned to a fine, sweet-smelling powder. It has transformed into a kind of mycorrizhae.

This too can be used as an inoculant tea, dynamized in water, the roots dipped in the solution before planting. Dynamization is often used in biodynamics when composing the preparations used in the field. Many nonpractitioners believe that the rules of biodynamics are in opposition to modern science. It's true that biodynamics does not rely on the lab for information or direction, but the process of dynamization is a tool that does come out of science. Dynamizaton, or actively stirring a liquid to create a vortex, then quickly changing direction to create an opposite vortex over a period of time, engenders a certain kind of molecular chaos that activates the elements within the tea or preparation.

The planting of vines is relatively easy, but the layout of a vineyard takes time. The most advantageous direction for the vines, the typical wind patterns, the slope of the land, and whether you will be using a tractor or horses or hand-hoeing must all be kept in mind when laying out the rows. The tractor part takes consideration. Tractors can be dangerous. The safest way to travel down the rows and turn around to go back up the row must be calculated. Often, some earthmoving is needed in a larger new vineyard to provide flat areas with enough room for a tractor to maneuver.

I adore horses, and the idea of tilling our land with one, but it is difficult for us to accommodate one here on the farm, or to find someone with a pulling horse who will come to help. While I believe heartily in doing everything by hand and in the old ways, I have also looked down the lengthy rows of the vineyard in the Champlain Valley that we lease. It is daunting to think of hoeing and scything that by hand. The raw reality is that we don't have enough hours in the day or the resources for a vineyard staff to do cultivation and mowing by hand. So I have come to appreciate the beauty of the tractor, and I am now enamored of them. I never knew that someday I would harbor jealousy in my heart for a woman I know whose husband bought her a vintage red FarmAll tractor for her birthday. In our vineyard, we don't have a riding tractor. We have a small vineyard and garden tractor that we walk behind, or, I should really say, Caleb walks behind. It is royal blue and was made in Italy. This tractor is Caleb's tractor. But I am eager for my own.

BIODYNAMIC COWPIT COMPOST

In our village, there is a gentleman who buys, refurbishes, and sells old tractors. In the summertime, he lines them up in the side yard next to his home, a phalanx of potentially marching venerable farmyard soldiers. I drive by this yard every day I go into work at the osteria. We slow the car down, and I long after the brightly painted red and shiny machines. I feel a little woozy.

A tractor is in my future. With the growth of our vineyard at home, and the possibility of new parcels to be planted in the next few years, a narrow-framed riding tractor is on the wish list. A vintage tractor it will not be, though. I must be practical. We must have something that is easy to repair, with parts easily procured, and all manner of attachments to help make the work of tilling and any spraying more efficient and reasonable.

Tractors of some kind, mechanical or equestrian, can be so important in the vineyard, and not just for planting, though they are necessary for beginning well. Sure, there are landscaping services that can come out and plant a new vineyard using laser devices and special machinery and their own tractors. They can set trellises and vines in one fell swoop. When planting a large vineyard this is often the best and most practical choice. But I like knowing that with our own tractor to prepare the ground, we can also do the layout the old-fashioned way, using measuring sticks and relying on the venerable Greek Pythagoras.

In a new vineyard, or with the addition of new planting blocks in an existing vineyard, new rows must be set. Before vines are planted in the ground, the width and length of the rows, and how much space exists between plants, must be decided. It is also a good idea to know how the vines will be trellised. Once these decisions have been made, the actual work of planting can begin.

BLAUFRÄNKISCH CUTTINGS

CUTTINGS AT THE HOME FARM

Trellising

Trellises are beautiful armatures. They bring a sculptural element to any garden, and they are necessary components in a more northern vineyard plot anywhere and in any size. Grape vines are climbing plants; they are eager for something to hold on to. I've always thought that the French word *grimpante*, which means "climbing" or "creeping," looks a little like a combination of "to grip" and "to run rampant"—the two things a grape vine is most happy to do.

PYTHAGORAS'S VINEYARD

In the old-fashioned way, start with the first row. Using a heavy-duty but malleable wire, make straight wire links in the length of the distance between plants. Fashion a loop on each end of the wire, and connect three pieces of wire together. This will be the measure. Place a post or stake at the first loop, then stakes at the second and third loop, holding them in a straight line. Do this all the way to the end of the first row, the last loop being a post or stake. Stand away, and line up the stakes by eye.

This is an abbreviated version. Friends of ours in France who own a vineyard link as many wires together as there are going to be plants in a row. They hold out the wire from the start of the row to the end, and place stakes at all the loops. This linking of straight wires would be a good wintertime endeavor to put together while watching a movie or sitting near the fireside on a cold snowy evening.

Then work on the sides. This is where dear Pythagoras comes in. In order to make sure the rows will be

PLANTING A NEW VINEYARD

parallel and the sides perpendicular, employ Pythagoras's Theorem: A2 + B2 = C2. This equation may seem daunting, but it is really very straightforward. I am not a math person per se, though I appreciate that the absoluteness of numbers has its own beauty.

In our home farm vineyard, our rows are 5 feet wide, and there are 4 feet between plants. Base row patterns on the machinery that will be used to keep grass down and for tilling, and keep in mind the trellising style. Rows must be as wide as a lawn mower, a scythe swing, or a pulling horse and contraption.

Our own Pythagorean equation looks like this: (5 x 5) + (4 x 4) = C2, or 25 + 16 = 41. Then turn to your calculator to figure out the hypotenuse. Surely you remember the hypotenuse from geometry class? It's the diagonal of the triangle created by two sides, in our case 5 feet and 4 feet. The length of the diagonal is the square root of 41. Plug in 41 multiplied by the √ on your calculator and, voilà, here's the length of the diagonal: 6.4 feet.

Once the length of the hypotenuse is known, make another straight wire at that length. Use that to keep parallels parallel and perpendiculars perpendicular. The best way to put this into operation is to measure that distance about every four rows using the corner posts and first stakes as markers. Do it in the center of the row, as well as at the end. This way things will keep even all the way along.

In a larger plantation of vines, tackle the placement of stakes first before getting plants settled. This is to avoid any backtracking and the need to correct posts or plants already in the ground.

IN ADDITION TO surviving by seed in the wild, transported by discerning birds and raccoons, grape vines will also grow and grow and grow until they are the largest plant they can be (there have been grape vines discovered that have arms as long as a mile). The largest living grape vine currently recorded in both stature and length is the Great Vine at Hampton Court Palace in England. This vine grows on an extension method where one plant fills a whole glass greenhouse. We would never grow a vine for production this way now, but the Victorians believed that this was the best way to get the most fruit. The vine is now 12 feet wide at the base, and the longest cordon is 120 feet in length. Usually it produces about 600 pounds of black dessert fruit a year, but in 2001 it had a bumper crop of 845 pounds.

The Great Vine was planted in 1769 under the auspices of Lancelot "Capability" Brown from a cutting taken from the *Vitis vinifera* variety Shiva Grossa or Black Hambourg at Valentines Mansion near Wanstead in Essex. Black Hambourg is also known as Muscat de Hambourg, or Moscato d'Ambourgo. It is the *vinifera* parent of our own La Crescent grape. In a Victorian manual on vines and vine culture, A. F. Barron wrote of several large vines that no longer exist except for the Great Vine at Hampton Court. In 1887, it already measured 4 feet around the base.

Although Queen Victoria never lived at Hampton Court Palace, the grapes from this vine were sent to the Royal Household at Windsor or to Osborne House on the Isle of Wight to grace the dining tables and picnic hampers. It was King Edward, who reigned from 1901 to 1910, who decided that the grapes were no longer needed by the Royal Household and could be sold to visitors.

They used to be offered in small wicker baskets out of St. Dunstans, a home for soldiers blinded in World War I. In World War II, German POWs

were given the annual task of thinning out the bunches of grapes, and people came from all over to buy the fruit from this spectacular vine.

Over time, the Great Vine has been housed in six or seven glass houses that have grown in size along with this stalwart plant. In the early 1900s a three-quarter span wooden glass house was built, which was a new shape and was designed on a different system than previous houses; it incorporated a viewing room.

In 1969, it became apparent that a new glass house would be needed. By that time, the Great Vine had become so entwined with the building itself that an aluminum-ribbed glass house was built on top of the old wooden one. The dormant vine was protected by a polythene sheeting while the old glass and the supporting frame were removed, leaving the iron structure of the 1900s and the Great Vine in place.

In the wild, grape vines will attach themselves to anything. In Roman times, farmers planted vines near oak or elm trees to allow them to curl and climb up the trees. This is the oldest form of trellising. The oak or elm and the vine had a symbiotic relationship. In Virgil's *Georgics*, which was written at some point between the years 29 and 37 BC, he speaks of this:

Land that tosses off the whiffs of mists and breathes in dew and

breathes it out again at will, and dresses always in its green attire,

and doesn't tarnish tools with rusts or marks

of minerals—that's the land for growing vines to trail

around your elms . . . .

These days, some two thousand years after Virgil wrote of this practice, we are slowly returning to this concept of trellising. As in everything, the pendulum swings from one notion to another over time. For years, trellising has been a regimented form designed to increase the yield of the fruit, and the ability to industrialize viticulture on a larger scale. Now, with more and more winegrowers focused on quality rather than quantity, work done by hand rather than by machine, the vision of viticulture returns to Virgil's time when our foremothers and -fathers actually knew something of the natural world, and were perhaps not so inured to it. As a sommelier, I see artisanal wine-growers explore older and older methods. As a hands-on winegrower myself, I am also intrigued by what these age-old methods might offer. Friends and mentors who grow wine in Austria, Werner and Angela, experiment with all kinds of ancient methods, and begin a vineyard planted with oak. Our friend Emmanuel, who grows organic wine in Burgundy, walks through our vineyard and along the hedgerow on a hot May morning, imagining a planting of Frontenac vines along the tree line growing up birch and maple next to the wild vines that have already taken root there. Not only is there a certain philosophical aesthetic at play here, but a curiosity about how the juice of the fruit that grows so wildly would be in the wine.

Already, Werner and Angela planted one vineyard according to principles of permaculture. They've pulled out some vines and established perennial garden beds of fruit trees and flowers to create more natural biosystems in their vineyard. Another vineyard was left *graupert*, which translates into "bed-head" or "messy hair" in honor of the feral look of the vines. In a Pinot Gris vineyard, they experimented with leaving established plants unpruned, in a semiwild state. They found that after three years, the plants began to regulate themselves and their production of fruit. The plants searched for their own balance. The wine made from these vines, also called Graupert, is easily one of the most intriguing and mesmerizing Pinot Gris I've ever had the pleasure to taste. They have begun the same process with one of their Pinot Noir parcels.

With cuttings of Frontenac waiting in the greenhouse to be planted this summer, we think of putting some along the hedgerow to grow "graupert" in our native trees, and to harvest their fruit (somehow before the hungry and savvy birds) to go into wine. Perhaps they will make a cuvée someday named Vergilius in honor of Virgil and his poem to agriculture.

Our home vineyard is trellised. We have experimented with a few different ways of approaching the best way to support the growth of the vines and their fruit. Certainly, for the cold-hardy varietals both Cornell University and the University of Minnesota, programs that specialize in cool-climate grapes, offer ideas about the best way to shape these plants. Their information is based on trial methods that result in good and plentiful fruit. They are research institutions, so their results are based on several years of experimentation and note taking. While cold-hardy varietals have been around for a while, the research is not exhaustive. Even now research like that of the Northern Grapes Project here in Vermont is being funded by sources like the USDA NIFA Specialty Crop Research Initiative to learn more about the cycles and habits of North American varietal crosses. How these plants grow

by personality is fairly obvious, but how they grow by place is still widely unknown.

Always, we have known that our trellising would have to be a certain height. The atmosphere of frost settles on the ground a few times a year, so it was clear that the plants would need to bear fruit relatively high. We looked at pergola trellising in northern Italy and understood that we would need something along the same lines given some of the similarities in our climates.

The very, very first vines we planted were ornamental and for cooking, vines that stretch up a pergola and onto a balcony at the back of our house. We planted them before we had any real thoughts about vineyards, so I don't count their history as part of our viticultural history, though perhaps that isn't fair. Those first three vines did, however, teach us about raccoons. For several years, we've had a family of raccoons nearby who thought they'd hit the jackpot upon arriving at our farm. Compost brought home from the osteria was the first casualty. Stale bread became the staff of their lives. We used to have an old truck with a sunroof that we had left open one evening by accident after a hot, summery night drive. In the front seat was a bag of old bread left over from work that the next day was destined to go out to the compost. We thought we were clever and outsmarting the raccoons by leaving it in the car overnight.

But raccoons love to climb, and they are very adventurous. The next morning, the bag had been opened and crumbs along with a remaining heel of bread were left in the two front seats of the truck. We could imagine our masked visitors sitting upright in the front seats and gnawing away all evening.

By the time of our first fruit on the pergola vines, we were so excited to see full, round black and white grapes, ripe for cooking in the next couple of days. But grapes are apparently the favorite food of raccoons, with stale bread a distant second. The grapes were gone before we could even think of picking them for a flatbread. The very particular claw marks on the sides of the pillars of the pergola left telling evidence.

By the time we were harvesting our own first vintage, the raccoons had become very comfortable dining at our table. During harvest, friends and neighbors come to volunteer in rotating shifts all day long, and we cook all day long to feed our hungry helpers. Our refrigerator is not nearly large enough for all the food we prepare and keep on hand. In Vermont style, we pack dishes up in coolers with ice packs or tie rope around soup pots and keep them in the cool shade of our porch or balcony. We place big stones on top of the coolers to ward off any unwanted diners. In the end, stones and rope do not discourage raccoons. We have come home to large pots of nettle soup mostly consumed, with raccoon paw prints stamped all over a porch bench that has particularly comfortable cushions, and a thick blanket with an almost completely round indentation of a slumbering guest. We've woken in the morning to the remains of a demolished apple tart, prepared the night before and locked in the blue cooler under a heavy stone. The stone was carefully laid aside, the top still on, the tart still in the cooler, yet all that was left was the perimeter of the dessert. It is assumed these raccoons don't really care for crust.

So between our resident raccoons with their very discerning palates and birds enamored of sweet grapes on the vine, the trellising must accommodate protection from such friends. While birds fly high and raccoons can climb, keeping the fruit higher does at least discourage the raccoons, and there is no easy way for the raccoons to get to the bunches except for those closest to the trellis posts.

Early on, during the second year of growth on our vines, I wanted to watch their natural growing habits and predilections before ultimately deciding on the trellising. We had planted the vines closely together, so I thought some kind of flat form of trellising would have to be used, but how

RACCOON'S NETTLE SOUP

many cordons, or fruiting arms, and what direction of shoot growth and height would still need to be decided.

I have seen and worked in vineyards with a single-arm Guyot system, which essentially means one arm going along the trellis wire in one direction. Many of the *vinifera* vines I've worked on have two to six buds or spurs that will produce fruit. I have also seen many double-arm Guyot trellises, with two arms going in opposite directions directly across from each other. I have also seen two arms set up in the same way with growth habits always going upward and attaching to wires above, a system that is called vertical shoot positioning (VSP).

But there are a plethora of trellising methods: There are high culture vines, pergola, gobelet, Pendelbogen, Sylvos, Cordon de Royat, Scott Henry, VSP, lyre, alberate, ballerina, basket training, Cassone Padovano, Casenave, eventail, Chateau Thierry, duplex, fan shape, Geneva double curtain, Lenz Moser, Mosel arch, Ruakura, Smart-Dyson, Tatura, and tendone. The list is somewhat overwhelming, but each system has its benefits, and the landscapes in which these systems were developed work with them in an interdependent way.

We monitored carefully the La Crescent and the Marquette. The La Crescent wanted to send a single arm out that would bend naturally to left or right with shoots coming off it. Marquette wanted to produce a bush or a tree with tons of fruit and foliage. The Riesling grows straight up, and the Blaufränkisch is still so small that it remains bushy. The blending grapes are all extremely vigorous, and given any opportunity will just lie down on the ground.

The vine is a creature of two sensibilities, two sides. One aspect of it wants to grow toward the sunlight, the other wants to send roots down deep into the earth. And while, ultimately, the plant will move toward sunlight, it also has a tendency to want to creep across the meadow floor. Its survival is dependent on both aspects. In the wild, if the vine is near a tree, it will use its gripping mechanism of tiny tendriled shoots to climb up the tree to the light. But if the vine finds itself in the middle of a meadow, just like all the little wild grape vine shoots we see every year hidden among the daisies, buttercups, vetch, bedstraw, and clover, it will move itself stealthily along the ground, sending out suckers to root in the soil to help support the ever-growing plant. Stupendous and uncanny, its prehensile, prehistoric, and sophisticated instincts have rendered the grape vine one of the oldest and most indomitable plants on our earth.

VINE TRAINED TO TRELLIS

GOBELET PRUNING

Grape vines grown for cultivation in our part of the world need to be given something to hang on to, in order to protect them from their predators, both botanical and animal.

To give new vines in the ground a little support with a wooden or bamboo stake helps them maintain a proper and upright position. Because vines want to crawl or climb, when left to their own devices without a tree, a stake, or a building to cling to, even baby vines will crawl and grow outward along the ground. They cannot support themselves vertically once they get to a certain height.

In the Northeast, the bamboo stake will last for a few seasons; it will eventually rot in the rain and the snow of winter. Wooden stakes or fiberglass are also options as our vines must grow higher for better ripening. We must keep them away from the ground where they can be damaged by early or late frosts, and away from animals that might be interested in such things as low-hanging fruit.

We began to trellis our La Crescent along wires first, as they so clearly wanted to wend off in one direction or the other, making an elegant march along their rows. It seemed easy to decide on a single- or double-arm Guyot system that would be relatively high, wires set between 4 and 5 feet. Whichever system chosen, the wires can never be higher than the width of the rows. This has to do with the trajectory of the sun across the sky and making sure the plants get maximum light during the day. Too much shade will be created if the wires are taller than the width. Too much shade means uneven or poor ripening.

With the Marquette, we thought to try a version of alberello or gobelet. *Alberello* means "small" tree in Italian; *gobelet* means "goblet-shaped" in French. We had just been to visit our friends Francesca and Margherita, who have a vineyard in Montalcino in Tuscany, and they have several blocks in Guyot, but one in gobelet. As we walked down the rows with Margherita, admiring the early spring bud break on the vines, she said that if she had to do it all over again she'd do the whole vineyard in gobelet. Gobelet has several advantages, the primary one being that there are no posts or wires to put into your vineyard, just one single hearty stake or narrow pole for each vine. Not only is it more economical, it makes it much easier to work in the vineyard. You can move around the plants easily, and chores like mowing and hoeing are more straightforwardly done.

Given that the Marquette showed a desire to grow in gobelet, we tried it out for a couple of seasons.

In terms of the growing shape of the plants, it worked beautifully; in terms of the fecundity of that growth and the pressure from fungal diseases that we have here in the Northeast, not as well. We found that despite our excellent air drainage on the farm, this form did not provide enough air circulation for the leaves and fruit, and plants that had a tendency toward black rot or mildews fell more readily. Marquette is more susceptible than the other varietals we have growing on the farm. So, the following year, we put in posts and wires, and began to convert the Marquette to a high-trellis double-arm Guyot.

French and American hybrids have what is called a semirecumbent attitude, meaning that the vines grow upward to a certain extent, but the arms like to grow out horizontally and a little bit downward. Forcing them into a completely upward form like vertical shoot positioning, which is what you see mostly in places like California, is not the most auspicious system for trellising these plants. Getting the trunks high, then allowing them to grow out from a certain height and fall like an umbrella, seems the most suited to their natures. We've adapted both cane and spur pruning methods to work with the vines that have this natural tendency.

Pruning, Simply

E ven though there are a number of different ways to trellis vines, there are two primary ways to prune: cane or spur pruning. When we plant young vines, we let them grow naturally their first season to establish their root system and their strength. Usually, we put in wooden or bamboo stakes behind them to keep them moving upward even in their planting year. In the spring of their second season, we prune the vine back to two buds. We monitor the growth of these buds to see which shoot develops as the strongest. We train the better looking of the two to the training stake. We don't remove the second shoot; we only cut it back to limit its growth. This is our "safety" or replacement spur for the next year if things go awry with the shoot we have chosen.

ONCE THE CHOSEN SHOOT becomes about a foot long, we tie it to the stake with baling twine or a fine plastic tubing string. While the plastic string is not biodegradable, it is gentler on the vine, easier to tie more snugly, and it traps less moisture close to the plant than the twine. Moisture in our environment can help breed disease, so we try to choose wisely in the materials we use when working with the vines, thinking for both the environment and the health of the vine. Ideally, we would have long willow reeds with which we could tie the plants. Our planting plan of native osier along the edge of the vineyard is not only to provide help with managing any underground water, but also to provide material generated on the farm with which we can tie the cordons every spring.

As the second-year shoot continues to grow throughout the season, we continue to tie it higher and higher up the stake, about one tie every 12 inches. This keeps the new trunk straight and from getting weighted down and breaking. Straight trunks are better for the vine: They produce fewer suckers at the base of the plant. We let the shoot grow to about 1½ feet above the top cordon trellising wire. At this stage, the vine gets cut back to just below the wire. This will encourage at least two lateral buds to develop and produce shoots that can become cordons.

It is important for the cordons to be as flat and as horizontal as possible. It is always better to have the cordon begin below the supporting wire and reach up to it rather than being forced downward with a hump. The hump can create inconsistent development of each of the buds on the cordon as the flow of the vine's life energies is broken up. When the cordon makes a gentle curve upward to become horizontal, the growth and vigor of each bud is much more even.

It is possible that the trellis wire is too high for the vine shoot to surpass in that second growing season. Not to worry: The training can continue in the following spring, starting over again from the two buds. It is always better to start from the beginning and a straight trunk rather than to prune a kink in the trunk midway. Trunks that are not straight tend to send out more watersprouts, which develop into hard-to-control vegetative vigor that diverts energy away from the core of the plant and the fruit. If the side shoots developed in that second season are not quite long or big enough to make true arms, cut the side shoots back to two buds on a spur in the spring and start over. Let them grow as much and as long as they can during this next season. If the vines are cane-pruned, a new cane for the cordon will be chosen again for the next season. If the vines are to be spur-pruned, the cane becomes the permanent cordon for the spurs.

To prune the established vineyard, I look at each plant individually, and I start by removing all the shoots and canes that I know will be not needed: watersprouts on the trunk, suckers near the base of the vine, and any dead-wood. Toss the prunings into the center of the row, away from under the vines, to be picked up later. Cleanliness in the vineyard is paramount. Leaving the prunings in the rows and simply mowing them and turning them

into the soil can work for an extremely disease-resistant vineyard, but when working organically in our part of the world, the chances are that prunings can still harbor overwintered mildews and black rot. It is best to remove them from the vineyard and burn them. Do not just simply add them to the compost unless they are deeply buried and the heat created during compost-ing can kill the pathogens.

For the prunings that are not chosen as cuttings for that season, we use a *brouette*, a burning barrow that the farmer who once owned our Vergennes vineyard built based on a design from our friends Tess and Emmanuel in Burgundy. The barrow is made from a 55-gallon drum barrel with doors that open and close on the top, dropped into a frame that has handles and wheels. As we prune, we toss the wood into the brouette, lighting the prunings on fire. We keep the fire going by constantly feeding it as we go along. On cold winter days this keeps us warm while pruning, and it burns the wood into ash that sifts through holes in the bottom of the barrow back into the soil. It is also perfect for cooking an impromptu snack of sausages with roast onions.

After removing the obvious, I step back and look at the vine. When cane pruning, I search for healthy canes that are growing on older fruiting canes relatively close to the trunk. The closer the cane is to the trunk, the stronger the arm. However, I steer clear from what are called bull canes—fat, extra-long, and vigorous specimens that outshine the rest of the canes. These tend to usually be unfruitful, because they are too focused on their vigor. However, they can be cut back as replacement spurs for the following year.

Once the canes for cordons have been chosen, I always leave one or two spurs below the cordon and as close to the trunk as possible. These are the "safeties" for next year. The safety spur should always point in the direction the cane should grow. Don't choose spurs that point outward away from the trellis. With the fruiting canes elected and the replacement spurs pruned, wrap the cordons around the trellis wire and tie with reed, twine, or that flexible tube tie.

The cordon arm length depends on the varietal and the age and health of the vines. For older, strong vines, choosing anywhere from ten to fifteen buds on each cordon can be a standard. I know some producers who choose eight to six. A little trick is to add an extra, then prune half the end bud to create a knob at the end of the arm. This provides a little bump to help keep the tie from slipping off the vine if need be.

Spur pruning works along the same concept but looks different. The cordon chosen in the second or third season has become a permanent arm along the trellis wire. In the first year the arm is positioned along the wire, the buds on the arm will form shoots. In the first season of spur pruning, cut that shoot down to one to three buds. With younger vines it's always advised to lower the number of buds so that the demands on the vine don't outstrip its strength. In the second season of spur pruning, there will be at least two shoots that developed from the buds. Choose the shoot that is situated closest to the cordon. Prune it down to two buds. Prune the other shoot or shoots to down below the lowest bud, thus preventing new shoots from developing from the spur. If the desire is to increase the crop, you can leave an additional spur with another two buds. Always choose the shoot that has the next closest bud to the cordon. The two spurs should form a *V* along the top of the arm. Adding spurs should be done judiciously.

Choosing spurs that face upward or spurs that focus downward is another choice to make while pruning, and, as always, there are various schools of thought on which method is better. Typically, spur shoots were always chosen facing up, and the lower buds were removed. But the weight of the canes can twist the arms of the cordon down and put an awful lot of stress on the plant if the crop is heavy, pointing the spurs downward anyway. In landscapes that have high summer heat and sun, it is usually better to choose a system where the spurs point upward so there will be an overhead canopy to shade the fruit from the intensity of the heat. In more northern areas with lower light, the spurs often face downward, as the leaves and fruit can catch more light this way to aid in successful ripening. Some growers use a combination, having plants alternate between the two forms. We currently cane-prune, so this is not a decision we have been faced with yet, but in our Champlain Valley vineyards, we are working toward spur pruning on some of the varietals. Even though our location is quite northerly, these vineyards receive a tremendous amount of sunshine, and the fruit is sometimes at risk of getting sunburned. Each site, no matter its latitude and longitude, will tell where to place the spurs.

Compost

Compost is the sacred substance of many gardeners and farmers. The magical transubstantiation of organic matter into soil. The old phrase from the Bible couldn't be truer: dust to dust. Everything returns to dirt.

WE'VE BEEN MAKING COMPOST SINCE the day we opened our doors at the bakery, before it became a restaurant. All the organic refuse from the kitchen came home to be composted and, once it had successfully gone through "the change," went back onto our small, first efforts at gardens.

Over the years, as our gardens have grown, so has our compost. It has gone from enclosures to heaps to windrows, and now some are back to enclosures. The young organic material from the kitchen is mighty enticing to the

HORSETAIL (*EQUISETUM*)

two yellow Labs that live down the road and make a quotidian pilgrimage. Last summer, of an early morning, I awoke to a black bear making his breakfast of the fresh leavings.

Bears and dogs and other visitors can actually be helpful, in that they dig up and turn the compost during their frantic sacking. But it is not necessarily something to encourage if you want happy neighbors, so our young food waste compost has gone back to an enclosure made with wooden pallets. The two Labs, Winnie and Oakley, are too good-natured to show their abject disappointment.

We also employ other composts. We have piles of horse manure from the farm down the road; we have cow manure from another farm that makes beautiful organic alpine cheese; and we have a composted mélange of manures, vegetal material, and farm mortalities, animals that have passed on and have returned to the earth. In the past, we have used the chicken manure from our neighbor's coop. That was before the fox finally took all but two old birds. The old hens retired to another farm, and the coop came down, the frustration of protecting the avian livestock from our woods full of coyotes, fisher cats, and foxes becoming too strong.

All these composts have their uses. Timing and type can be everything. Cow manure aged for three months is best for the vineyard. A little horse manure in addition can also be a good thing. The farm mortality manure is excellent for raised flower and vegetable beds; our own organic material compost provides a shot in the stem to the vegetables as well. Chicken manure is perfect for roses. Or a slurried, liquid compost tea made from stinging nettles.

But these are very straightforward ways of thinking about compost when in truth composting is a rather arcane art, and how could it not be compared to magic, or alchemy, when you see one thing change from one state to another—not unlike the creation of mycorrhizae in a buried cow horn.

In biodynamics, soil is key, but compost is the cornerstone of farming. The creation of compost using fermented inoculants is critical to developing compost faster, giving it greater nutrient content and better tilth, or crumbly structure. Plus, this kind of compost is less inclined to produce a stink.

In biodynamics, there are six compost additives; they are composed of plant parts or extracts treated with animal tissues and buried to aid decomposition. Five of the decomposed preparations are humuslike, with very friable texture; the sixth is liquid. The main ingredients of the preparations are yarrow, chamomile blossoms, stinging nettle, oakbark, dandelion, and valerian extract. These preps are truly not unlike the culturing of mycorrhizal fungi, in that they are stuffed into various animal tissues like stag bladders, cow mesenteries, and skulls.

Scoffers and skeptics abound when you start to talk about burying plants in animal parts for six months. Witchcraft and paganism is on the tip of the tongue. But these practices are not without official and scientific examination. Not only are there several organizations and research farms in Europe, such as Maria Thun's in Germany, that supply information about the efficacy of composting in this way, but I always like to point out that a thorough investigation of the traditional organic compost versus the biodynamic compost was done by our own USDA over ten years ago; it surprises me that it is almost never referenced.

BIODYNAMIC COMPOST

IN 2000, THE PUBLICATION Biological Agriculture and Horticulture published a USDA study titled "Effects of Biodynamic Preparations on Compost Development." The study compared biodynamic, or BD-treated, compost and control compost in raw material made of dairy barn waste, consisting of manure and pine shavings bedding from the Washington State University Dairy Center. Washington State University has one of the most progressive horticulture and viticulture programs in the country.

There is plenty of previous research that has found that BD preparations can speed up composting and provide a higher quality of compost. Compost treated with the BD preps also showed a higher carbon to nitrogen (C:N) ratio, and produced less ammonia and more nitrate. Straw treated

with all six of the BD preps released more carbon dioxide rather than meth-ane, suggesting more decomposition in one year than compost not treated. It turns out that methane is several times more damaging to the environ-ment than carbon dioxide. Higher composting temperatures are consistently reported in BD-treated composts. Maturing BD compost does not always reach the peak temperatures of those composts not treated with the preps, but the BD piles retain the temperatures longer. In several composting meth-ods, with and without BD compost preps, the BD-treated composts consis-tently had higher cation exchange capacity (CEC) per unit of organic matter.

In the results of the USDA study, they found many of these precepts to be true. Temperature in BD compost was higher than in the control, and a higher temperature was a consistent effect of all the BD-treated piles. Higher temperatures suggest more microbial activity, which can lead to faster matu-ration of compost and a greater reduction of pathogens.

The average pH of the piles stayed relatively consistent between BD and control composts. However, near the end of active composting, the pH in the BD compost dropped temporarily, giving them a more neutral pH. Potential enzymatic and biological activity is generally greater at a more neutral pH. At a lower pH, compost is also less likely to lose nitrogen through ammonia volatilization. Nitrogen content in the final compost piles was consistently 65 percent greater in the BD-treated composts. In terms of microbial activ-ity, finished BD composts were higher in bacteria and lower in fungi.

Several measurements reported in this study supported that BD compost-ing was in fact different than non-BD-treated compost. BD composts were truly hotter—higher temperatures indicating greater microbial activity, which generally results in faster decomposition and better control of weeds. Treatments that increase the temperature may result in a better processed compost, or a better finished compost.

One of the biggest points of contention between biodynamic practi-tioners and non-practitioners is that such small amounts of preparations are used to treat raw manure. Non-practitioners never believe such small amounts could have any effect. Yet we know that the effect of many bioac-tive compounds (substances that have direct effects on living organisms) do affect the growth of plants and are most effective in trace amounts. Effects of the BD preparations could also be caused by gaseous or liquid chemical factors. Yarrow, chamomile, nettle, and valerian are well known as medicinal plants and contain a variety of what we call bioactive compounds. Extracts of chamomile, for example, have antibacterial and antifungal properties.

These or other biologically active substances, which act at very low concentrations, might be present in the BD preparations and may be the cause of the effects seen.

I am left wondering, given the long history of research using these compost preparations, and taken in light of the USDA report, why doesn't everyone practicing organic farming prepare their compost in this way?

For many, the BD inoculation process may seem overwhelming when you already have a long list of chores to complete. But upon closer examination, to set up your compost this way is actually fairly easy.

In Maria Thun's writings, composting biodynamically is pretty simple. Make a heap of compost material mixed with straw. Make five holes about 20 inches in depth, two on each side, but toward the top of the heap, and one in the center of the top of the material. Insert the five basic preparations: dandelion, yarrow, stinging nettle, chamomile, and oakbark. Water the whole heap well with the valerian preparation.

At the Josephine Porter Institute here in the United States, they recommend the Pfeiffer program, which is ultimately even simpler. It consists of a starter packet that contains the six preps as well as some of the horn preparation. The starter gets mixed with enough lukewarm water to make a moist paste. Store the mixture in a warm place for about twelve hours before application. Dilute the paste again with more water, enough to fully wet the material to be composted. The amount of water is dependent on the spray equipment or the method of application. Apply the mixture while building the compost pile, thoroughly mixing the starter in with the material.

We know these inoculants can cause increased bioactivity, but how do these individual plants work as plant medicine or boosters? What botanical information can we glean that will help us understand their role in making compost, and their role in relationship to the plants they will treat?

All living plants need trace amounts of certain minerals: potassium, calcium, magnesium, silicic acid. The six plants used in the BD preparations all have elements that stimulate the plant's ability to absorb these minerals. Nature works in amazing ways. In biodynamic thinking, it is accepted that the world is no longer operating to the purest of its ability. Too many modern, industrial, and agricultural factors have affected our natural world. Many soils are overused, polluted, or given filler rather than real nutrition. Using straight composted manure may no longer be enough to encourage growth. Manure inoculated with the plant medicine seems to help the compost work more effectively.

Teas, Concoctions, Sprays

Approaching compost and manure biodynamically naturally leads to the use of plant medicines. Those six primary plant-derived inoculants for compost are the same core plant medicines used in the garden and on the farm, in addition to the very effective and important horsetail tea. They can be used as preparations treated with the animal tissues, or as decoctions, infusions, or prolonged infusions, also called slurries.

FOR THE PAST THREE YEARS we have used a variety of teas on our plants and in the vineyard. We have focused on chamomile, yarrow, nettle, and horsetail. We have also sprayed the horn manure at the beginning of the season, and another preparation called horn silica when we have had too much rain, and to bolster the ripening.

Two years ago, on a sunny early spring day, friends who have a farm just a bit down the road held a big party and a fundraiser for the local Waldorf School. The Waldorf methodology is part of the Rudolf Steiner curriculum, so it makes perfect sense that as farmers they would be interested in the study and implementation of biodynamics as well.

As part of the festivities, Joseph wanted to get a group together to make a preparation called Cow Pit Pat compost, or barrel compost that is sprayed onto your soil as a compost tea and is considered a jump-starter for the land in the spring. To make barrel compost, a group of people willing to work hard for an hour is needed. A pile of fresh cow manure gets chopped up with eggshells and basalt and flipped repeatedly with shovels, kind of like kneading bread. This prepares the manure by breaking it down enough before it gets buried in a barrel in the ground.

There were probably about ten of us, and we worked in a circle. When one person got tired, another would sub in while each of us took a break. It is backbreaking work and would be grueling to do by yourself or with just two

THE NATURE OF MANURE

In the vineyard, there are essentially four different manures that you might use or mix depending on your location and the characteristics of your land. In biodynamic agriculture, when we look at the vine's relationship to animals, we must approach the animal world in the same way we approach the world of vegetation. It is helpful to think about the temperament and nature of the animal just like we examine the botany of a plant and to contemplate how that might be reflected in its manure.

Horses, for example, are often described with natures of heat. They are sensitive and nervous creatures who can respond abruptly to something they see out of the corner of their eye. A horse might skitter sideways, or buck to unseat a rider. They like to run and jump. They are animals that like to take to the air. A cow spooked would most likely do nothing, or it might round its back and dig in its heels, trying to dig its way further into the earth. A cow also has a connection to liquid. They are dominated by the production of their milk, and their manure itself is more fluid. The cow has the longest intestine of vertebrate animals. Goats are animals with horns—their attention is raised toward the outer world, their horns are antennae for a variety of information that help them survive in the wild. A wild goat also needs very little water to survive, suggesting it's an animal that responds to light and the effects of the sun. Goats like to climb like horses like to jump. They will stand tall on their back two legs in order to get to the soft green leaves in a tree. Pigs are connected to the earth. They use their noses to literally root out food.

So the manures of these animals can be used in different ways. In the biodynamic paradigm, the more liquid manure of the cow is good for leaves and foliar development; the manure of the goat, an animal of light, acts on the flowering of the grapes and as a consequence has some effect on taste; the hotter manure of the horse is beneficial for fruit development and ultimately for flavor. Horse manure would also benefit vineyards in northern areas, providing more heat. Pig manure would specifically give a dose of healthy food to the roots of young or struggling weak vines.

Nicolas Joly writes in his book *Wine from Sky to Earth*, "On the same patch of vineyard, when the

vineyard soil is alive and the plants are healthy, using four different types of manure can lead to four different types of wine made from these grapes." It would stand to reason then that the plants would respond to the different qualities of each of these manures and, as a consequence, the winegrower needs to think carefully and choose wisely the kind of manure she might use to aid her plants. Talk to older winegrowers: the ones who plow their land with horse-drawn plows, or who farm on such steep, Roman-made terraced hillsides that they must do everything by hand; the ones who live the life of their land in their bodies and blood. These winegrowers talk about the vines having a sensitivity. They like to use the horses to plow because they feel the vineyard responds to the nature and energy of the horse. These farmers, just like all the farmers before them, have developed a keen sensitivity themselves, and a respect toward nature—something that can't be taught in agriculture schools or understood by intellect alone.

We, ourselves, now use a combination of cow and horse manure in our vineyard. Because our climate is prone to issues with fungi, we hope to make the soil healthier so that the pathogens stay in balance, meaning that they remain in the soil rather than jumping to a new host plant. The cow manure strengthens the leaves of the plant, making them more resistant to disease. The horse manure adds that bit of heat to our northern aspect, aiding in the fruit development, and eventually in the ripening and taste.

We have also found that a useful way of addressing some of the soil and vine needs can be accomplished through compost tea. The two vineyards we lease and work in the Champlain Valley are much larger than our home vineyard; they cover a lot more ground with wider rows and more space between plants. The timing of our reclamation of these vineyards coupled with their size has not given us the opportunity to grow or acquire enough compost to effectively start to change the balance in the soil yet through an intensive composting program. We are relying on cover crops to help us in the early stages. We also use compost tea made from the organic cow manure compost from down the road and believe it will go a long way toward helping us strengthen the soil and the vines, while at the same time acting as a natural suppressant to our host of fungal diseases. The tea-brewing process elicits and grows beneficial bacteria and fungi and then suspends them in the water in a form that makes them quickly available to plants. We wouldn't rely on this every year as a way to handle soil fertility, but as an interim measure, or as an addition to full-on compost soil amendments, this can be a good way to get started and to help keep the health of the vineyard moving forward.

To study and make use of ancient knowledge in agriculture—seeing how to improve it according to the ever-changing needs of the land—is often laughed at by modern farmers fresh out of agriculture school who have been taught that the only way forward is in chemical and large-scale farming. Plants and animals are treated as only food for that season. There is very little thought put into the future of these farms, though these new and "improved" systems are often touted as the future of agriculture. In a vineyard, just like on any other farm, you must look three years ahead. A farmer must ask himself, "Are the choices I am making now going to improve and sustain the quality of the soil and the health of the plants or animals three vintages away?" If you are tied up only in the concerns of how much you can harvest this year, you cannot be thinking ahead.

Discussing compost and manure leads to thoughts of the sustenance of the farm. Compost and manure occur on ground level. The energy and nutrients they deliver to the soil, to the roots of the plants, to leaves and flowers affect a whole biosystem of life on the farm, and are crucial to the safeguarding and husbanding of that biosystem for years to come.

people. All of us involved in preparing the compost would be able to partake, sprinkling it onto our own land.

Once the compost is adequately mixed for that hour, it gets layered into a wooden barrel without a bottom and buried in the ground with a wooden lid. A layer of worked manure gets put down, along with the same five preparations in five holes just like in the regular BD-treated compost, then it's all sprinkled with valerian extract. Another layer of manure gets added, then the preps, then a sprinkle again with valerian, and so on. After four weeks of sitting in the ground, the barrel is opened up. The manure is taken out and mixed up, then it gets layered back in with the preps and valerian once again. After another four weeks, it's ready to use and to be sprayed as a compost tea on the land.

This process reminds me of other fermentation methods. I think of traditional Korean kimchee buried in the ground in clay pots, and I cannot forget the large traditional clay amphorae that the ancient Greeks and Romans buried in the ground to ferment wine, or the even older beeswax-lined Georgian clay *kvevri* that are big enough to stand in with your arms outstretched. The fertile valleys of what we now call Georgia in the Caucasus are believed to be the first home to cultivated grape vines and Neolithic winemaking over eight thousand years ago.

Barrel compost provides an extra boost to the farm and the regular biodynamic compost. It heightens the activity in the soil. It is particularly useful for new farms transitioning from long-term conventional use to biodynamic methods. It also appears to be very beneficial for soil that has been contaminated by radiation. As part of a research experiment, it was found that, after the Chernobyl nuclear disaster, soils treated with barrel compost rebounded more readily from the radioactivity.

Using herbal teas based on the biodynamic paradigm is a way to use plant medicine that enhances the life of the soil and the strength of the plants instead of resorting to chemicals that will strip the soil of life and ultimately compromise the heartiness of the plant. Chemical agriculture follows a specific cycle. Pesticides developed from chemicals may prove effective for a time on the pathogens or pests disturbing the plants, but those pesticides, when used regularly, weaken soil and plant health, so that before long fertilizers must be applied to ensure the plants' appearance and production. Proponents of chemical agriculture compare this to the need of the organic farmer to use compost to aid the soil; they say it's the same cycle of depletion and replenishment. However, the chemical fertilizers do naught for the soil or for the long-term life of perennial plants like vines. They do not work

HORSETAIL AND NETTLE COLLECTED FOR BIODYNAMIC TEA

below the surface level of the soil, nor do they encourage the web of life in the soil. They do nothing but fill the plants with false food, junk food, nothing tangible or useful to them in the future. These fertilizer fillers ask the plants to grow so fast and be so top-heavy that they can't support their own structure down in the roots.

Before there were chemicals, farmers used plant medicines for thousands of years, and there is much knowledge in the little bits of plant wisdom left to us in places like *The Old Farmer's Almanac* and vintage books like *Old Wives' Lore for Gardeners*. This is the same stuff that biodynamics is based upon.

Herbal teas for the vineyard focus on certain elements necessary for vines to be strong and produce good-quality fruit. Field horsetail, the *Equisetum arvense*, a prehistoric fern that tends to grow in wet areas, is a natural fungicide. It is made primarily of silica, which attracts sunlight and offers a drying action on the soil. Fungi grow in wet and humid circumstances that occur at warmer temperatures during the summer months, so medicines that can dry out that humidity are very helpful.

Stinging nettle is a good foliar spray, strengthening the iron and magnesium elements in the leaves. It improves sap circulation. Yarrow tea increases the potassium in the plant; it has trace elements of sulfur, which is one of the primary needs of plants. Chamomile adds calcium. Dandelion blossom tea is another plant tea high in silica, helping to strengthen the leaves of the plant against foliar diseases. Parasites find it harder to penetrate the leaves, especially leaves pummeled by constant wet weather. Valerian blossom tea is used sparingly and is historically recorded as being sprayed on the ground around all fruiting plants during or near St. John's Day—the Summer Solstice. In vineyards, it's believed to help in bud formation for the following year. It's also used to help soften the effects of frost, as it creates a warming effect within the plant.

Many biodynamic practitioners are stringent about following the set protocol of sprays and composting every season. Some farmers who follow the biodynamic strictures for a period of time find that their vineyards become so healthy and strong that they eventually pursue a modified course and don't use the plant medicines or composts if they don't actually need them. I know farmers who have managed their land and vines biodynamically for a period of time then tapered off to fewer to no biodynamic inputs because the soil and plants had reached a balance. Their farms tend to also be blessed with a near-perfect climate for growing wine: sunny and dry in the summer, and just cold enough in the winter.

Here in Vermont, our climatic conditions are perfect for the proliferation of three of the most detrimental diseases: black rot, downy mildew, and powdery mildew. We experience the same weather conditions that are perfect for the diseases anthracnose and phomopsis as well, but those are less pronounced than what I think of as the Fungus Triumvirate.

In many other parts of the world known for winegrowing, there are similar atmospheric conditions, and those winegrowers confront the same issues as we do every year. Because these fungi are scourges and can decimate a crop if not a vineyard, both organic and biodynamic farming do often rely on two mineral rather than plant-based preparations: sulfur and copper.

In both conventional and organic agriculture systems, sulfur is considered essential for healthy plant growth. Just like nitrogen. Just like oxygen. Just like hydrogen. Just like carbon. These five elements are crucial to the way plants make their way in the world. Sulfur is the connector among them all. In biodynamic thinking, sulfur is compared to the mediator, the substance that makes the plant ready and able to accept these other beneficial elements.

Sulfur is found in the amino acids that make up plant proteins, and plant and animal scientists have shown that plant tissue should contain one part sulfur for every fifteen to twenty parts nitrogen for optimum growth and production. Sulfur is active in the conversion of inorganic nitrogen into protein. Sulfur is a catalyst in chlorophyll production, and it makes possible the formation of nitrogen-fixing nodules in leguminous plants (think cover crops). Sulfur is also a component in various enzymes. Like nitrogen, sulfur is a very mobile and flexible nutrient that can move rapidly downward through the soil, and sulfur counts, just like nitrogen counts, tend to be high in soils rich in organic matter. But under intensive cash-crop rotation and production, the breakdown of the sulfur inherent in the organic matter may not be fast enough to feed the plants the nutrients needed to keep up with the demands of high yields.

Elemental sulfur must be converted in the soil to sulfate in order to be available to the plants. This conversion is performed by soil microbes and requires soil conditions that are moist, warm, and well drained in order to happen in a timely manner. In the right circumstances sulfur can move through the soil 50 percent faster than nitrogen. Even in clay-intensive subsoils, it is common to find that sulfate that has leached through the soil over a period of time is "perched" on the clay layer and available to plants once their roots reach this zone of the soil.

Animal manures and sulfur-based fungicides are a way to supply sulfur to plants. Elemental sulfur and lime sulfur are two of the oldest known pest control materials and have for a long time been widely used as very effective fungicides that also have very beneficial effects for the plants and soil. Homer wrote of sulfur's "pest-averting" qualities in the ninth century BC. Residues from spraying and dusting sulfur go into the soil, and rather than being harmful to the person eating the harvest on the plate, this application provides the plants with the sulfur needed for photosynthesis and fruit formation.

In organic or biodynamic methods, when using sulfur as a fungicide, it is used in very small amounts and provides those trace amounts that are necessary in developing a good, healthy soil.

Here, at the home farm, we spray sulfur more toward the beginning of the season and through blossoming, then one or two more times throughout the season depending on the weather patterns. Sulfur is particularly good for combating the two mildews, downy and powdery, and when started after the first five leaves on a plant show themselves, in a cycle of spraying every ten to fourteen days, it helps keep pathogens at bay. Sulfur is never curative if the mildews attack, but it can control the proliferation and lessen the damage.

HOW TO PREPARE PLANT TEAS

Tea sprays can be prepared easily. The general procedure for making plants into teas to spray in the vineyard, garden, or orchard is just like steeping tea to drink.

To make flower blossom teas from yarrow, dandelion, and valerian, add 1 gram of dried blossoms to 10 liters of already boiled water. The boiled water is then poured over the blossoms. Let steep for 15 minutes, then strain through a sieve. The tea must be cold before putting it in your sprayer and applying to plants.

For stinging nettle, add 5 grams of dried leaves to 10 liters of water, boil briefly, then let steep for 15 minutes before straining. Allow to cool before spraying. With nettle, you can also make a much stronger tea with 5 grams boiled in 1 liter of water, then dilute it by adding another 9 liters of water later.

Horsetail and oakbark must be added to cold water, then boiled for 15 minutes, strained, and cooled. One gram of horsetail and 5 grams of oakbark are used for every 10 liters of water. Once the tea is cold, spray in a fine spray.

Ten liters of tea are needed to cover a little over ½ acre of vineyard or garden space.

HORSETAIL DECOCTION

NETTLE TEA

Copper has another history. Copper micronutrients are essential to human health. In agriculture, copper salts or metals are used in various forms: sulfate, hydroxide, oxychloride, and cuprous oxide. In viticulture, copper can be used alone or in combination with lime sulfur. Because we use the sulfur separately, we use a copper hydroxide to also combat fungal pathogens, in particular black rot, which can be very intense on our hillside under the right weather conditions.

Copper hydroxide has been known since copper smelting began around 5000 BC, although alchemists were probably the first to actually manufacture it for use. This was done by mixing lye (potassium hydroxide) and blue vitriol (copper sulfate), both of which were known in antiquity. It was produced on a large scale during the seventeenth and eighteenth centuries for pigments such as blue verditer and Bremen green, colors used in both ceramic glazes and painting.

Copper hydroxide occurs naturally as a component of several copper minerals: azurite, malachite, antlerite, brochantite. Copper hydroxide is the primary ingredient in what is known as Bordeaux mixture, the classic copper fungicide used in vineyards since the nineteenth century when the importation of American vine stock to Europe brought not only the Great Wine Blight of phylloxera, but also a host of mildew fungi.

Copper achieves its effects as a fungicide by means of the copper ions that respond to the enzymes in the fungal spores in such a way as to prevent germination. It is best used as a preventive measure, but can also be used to control further fungal proliferation once a fungus has gotten a foothold. It does not kill the fungus of black rot—in fact, no organic or chemical treatment currently produced really can. Treatments can only prevent and sublimate. Even harsh chemical fungicides will only break down the fungus by dissolving its cell walls for a short period of time; then the fungus, wily as only nature makes living things, will morph and change in order to resist the treatment. Because of the way copper ions interact with the enzymes, the pathogens don't change and adapt to the treatment, but you must begin to know the pathogens as intimately as you know your plants in order to manage your spray schedule in sync with the cycles.

Copper and sulfur, when used inappropriately—and by this I mean sprayed too frequently, too heavily, and for too long—can have a negative effect on the land and on sensitive plants. Just like any useful thing used excessively, they can become toxic. Copper is considered a heavy metal and if used heavily, it can decrease if not decimate an earthworm population, as well as other beneficial organisms in the soil and on the plants. In both organic and biodynamic philosophies, the recommended dosages of copper and sulfur provided on the back of the store-bought bags are considered too high. But when used in smaller doses than usually recommended, sometimes only a tenth of what is recommended on the bag, and mixed along with plant teas like horsetail and nettle, you can get the best of their benefits and also regulate how they interact with your soil. In France, a recent study showed that farms worked organically and biodynamically with minimal copper and sulfur inputs showed no degradation of soil microorganisms.

The more I understand, as I learn the cycles of the plant, animal, and microbial life of my own vineyard, the more I realize that, if the work is done properly at the beginning, by attending to your soil, the vineyard will find its own balance. Pathogens will be more inclined to remain in the soil, and nutrients will feed the plants. With thoughtful and minimal use of the copper and sulfur minerals, the soil can still thrive.

Pesticide is a strange word. Growing up, I always thought it meant something that would eliminate insects or rodents. I never realized until I began caring for my own vineyard that it was more encompassing: Pesticide is a treatment that affects insects, animals, and pathogens. It is a difficult word, because it conjures the image of dangerous toxins in order to manage a problem.

In biodynamic thinking, problems don't exist in the same way we tend to approach them in the conventional world. There does not have to be panic and anxiety, reactions rather than responses. I've found that, conceptually, biodynamics, just as all traditionally based farming systems, is more akin to Buddhism than to modern Western dialectic thinking. If a problem arises on the land, biodynamics recognizes that there is an imbalance at the core rather than a problem that exists in a vacuum. While there are tools to help mitigate these problems, the biodynamic farmer will return to the soil and begin to examine the ways in which the soil, flora, and fauna can be better managed. The problem exists at the moment, but it won't stay the same. It will shift and change.

In farming, there will always be insects and fungi. There is no way to eradicate them, nor should there be. Each bug or fungus has a role in the delicate ecology of our natural world. It is easy to hate something like the Japanese beetle. I myself have harbored much ill will and frustration toward this lustrous scarab, but the Japanese beetle is only a scourge because it is not indigenous to our environment. It traveled here on plants from eastern Asia, plants like roses, and has no natural predator here in our Western world. Just like the fungi and phylloxera traveled to Europe on our American grape vines and caused a huge plague in the nineteenth century. The intro-

BORDEAUX MIXTURE

After downy mildew was first noted in France in the late nineteenth century, professor of botany Pierre-Marie-Alexis Millardet of the University of Bordeaux studied the disease. In his studies, he noticed that the vines that were planted next to roads were not affected by the mildew, while the ones in the center of the vineyard were the most badly damaged. After making a series of inquiries, he learned that the vines along the roadside had been sprayed with a mixture of copper and lime to deter passersby from eating the grapes— the treatment was both visible and bitter-tasting. This chance discovery led Millardet to begin experimenting with the treatment. He did the majority of his work at the Chateau Dauzac in Bordeaux along with Ernest David, Dauzac's technical director. He published his findings in 1885 and recommended the copper-lime mixture to combat the by then endemic downy mildew. Continued research has shown that Bordeaux mixture is effective against a whole battalion of fungi that attack vines and other plants.

duction of these organisms disturbs the balance of native ecology, yet we have the ability to help restore the balance by focusing on the health of the soil and raising stronger, more naturally resistant plants.

The same holds true for our native pathogens. The mildews and black rot belong to us here in the northern corridor, and when other elements in our agriculture are out of equilibrium, so too is our plants' innate ability to resist these diseases. Cool-climate hybrids have a good dose of indigenous wild grape vine in their DNA (the wild grape species being much more resistant to these diseases as their evolution developed side by side with the native pathogens), even though they also have a large percentage of *vinifera* in their lineage, a species not resistant to these culprits. Another conventional winegrower recently commented to me that he was envious of our cold-hardy varietals because of their built-in resistance. We have better chances of success growing organically. He feels he does not have that luxury himself, growing only *vinifera*, non-native vines that are more susceptible to native pathogens. Just like the heirloom roses from places like England and France that we have planted in our rose garden, these plants fall easy prey to black rot. Whereas the wild native roses that I have planted here are blemish-free.

Pests of every stripe will abound if keen attention is not paid to the garden. The insects that find their homes here, as with all else, find it in our soil. Grape flea beetle, rose chafer, Japanese beetle, tumid gallmaker, even phylloxera. They are all part of the living web of microorganisms moving through our silty, loamy, clayey, or sandy earth.

This past year was the first season I saw such an insect pest called the grape flea beetle, but it may only be the first season that I noticed it, and my attention was more directed elsewhere in spring and summers past. My attention has a learning curve, and it grows with each year that I work with the land. The grape flea beetle is a very pretty native as far as beetles go: very petite with a black shiny armor that is iridescent with green and blue. Sometimes it is called a steely beetle. He is about the size of a coarsely ground peppercorn, and hibernates in a forest-edge ecology: wastelands, woodlands, or abandoned vineyards. In the spring, overwintered adults migrate to the grape vines at about the same time as the buds begin to swell. They bore into the nascent buds, hollowing them out. I've also seen them feed on leaves, as do the wormy dark-colored larvae they leave behind.

Our block planted with La Crescent is right on the edge of a tree-lined stream, a narrow forest that separates our meadow from the neighbors. The edge where the leaves fall provides a perfect hibernating ground for the

GRAPE FLEA BEETLE

GRAPE CLUSTER WITH DIATOMACEOUS EARTH

grape flea beetle. So do the grape vines in that section, if they are not cleaned up properly in the autumn after harvest. Cleanliness is next to godliness in the vineyard. To keep these beetles at bay, cleanup is paramount, as is cultivation. Hoeing and cultivating under vines for organic growing is important in many places for many reasons, and this is just another. Relatively frequent discing in the spring between rows will lightly break up the soil and the pupal cells residing therein. This exposes the fragile pupae and dessicates them.

Once the beetles are seen, though, another measure can be employed as well. Sprinkling the plants with diatomaceous earth provides very good natural control. Diatomaceous earth is a naturally occurring, soft, siliceous sedimentary rock that is easily crumbled into a fine ivory powder. Its chemical composition is about 90 percent silica, with trace percentages of alumina, a clay mineral, and iron oxide. Diatomaceous earth is made up of the fossilized remains of diatoms, a type of hard-shelled algae. It has sharp, cutting edges, and it works as an abrasive on insects that have an exoskeleton, and as a drying agent. The fact that it comprises mostly silica makes it a great friend in the vineyard, as it also reflects sunlight.

We found that this natural treatment also helped immensely in controlling the various mildews and black rot in what was a very wet season. This added benefit gave rise to the question of a relationship that might occur between the life cycle of the grape flea beetle and fungal pathogens. I wonder if the environmental conditions in which each flourish are connected. I only have one season with the grape flea beetle to refer to, but the apparent result begs for experimentation and monitoring. This is the way the farmer must think: looking for connections, posing questions, responding to the events that occur in the environment of the farm.

We often make jokes about diatomaceous earth, not because it's inherently funny, but because the name makes it sound stupendous, as though "diatomaceous" were some kind of hip vernacular for the best of the best. As it turns out, it can be stupendous, and it does a quick job if the application is followed by forty-eight hours of dry, clear weather.

We see the occasional rose chafer, usually in our blocks with sandier soil. She is a North American beetle of the family Scarabaeidae with a pale beige body, sienna-colored legs, and a black dorsal stripe. She undergoes a complete metamorphosis during development, after having overwintered deep in the soil as a white-bodied larva. Once rose chafers have transformed from their pupal state, adults live for three to six weeks, feeding on bright green plant material and indulging in mating. They lay their eggs in sandy soils, and new larvae will feed on the roots of grasses and weeds, or ornamental plants if they are in the flower garden. Their mastication of the leaves of plants releases a volatile compound from the plant that acts as a beacon to other rose chafers. They come to swarm and in the vineyard can eat both leaves and fruit buds, skeletonizing leaves and destroying fruit. They are called rose chafers because they have a particular love of dining on rose leaves.

Given that we have both a rose garden and a vineyard cheek by jowl, we are lucky to have a more loamy, silty, and clayey soil that is not hospitable to the chafer. The sandier section of our land, which is nearer the brook, doesn't seem to have harbored many specimens of this pest. Control of the chafer is the same as for the flea beetle. Good cultivation and a little diatomaceous earth go a long way.

Japanese beetles always arrive around the Fourth of July. There is something suspicious about the fact that they make their presence known on a holiday, as if they delight in catching you unaware, when you have relaxed for just a moment, let down your guard, and raised your glass to the bounty of the upcoming season. June is one of my favorite months. Everything is blooming and everything looks like the paradigm of health. The roses are in profusion, the leaves are in full green, the vines have just set their fruit. The apples, small and hard, are just starting to blush in the hopeful heat. Then we pass into July, and it's as if all the pests and their minions storm the walls of the farm, bent on destruction. I think of the Japanese beetle as their leader. Large, well armored, and seemingly dumb, they appear slowly, just one here, a few over there, nothing to worry about. Then all of a sudden they have flocked, swarmed, thronged, clustered, herded, and arranged themselves in a multitude and made a fine fretwork of leaves in the vineyard and rose garden alike. The leaves have an absurd beauty in their lacelike condition, but I know full well that without leaves for photosynthesis it is just a matter of days before the plants might wither and die.

I am at my most vituperative when I speak, think, or write about Japanese beetles. I counsel myself to be more Zen in my thinking. They have yet to destroy everything, and the plants have never actually rolled up their roots. Two years ago, we had a particularly bad infestation. The year before that I thought it would be enough to walk around with a jar of soapy water and drown them. But clearly this was a fool's errand. Once they are left alone to proliferate, they squat in numbers and ravage every plant in sight.

In my more Buddhist moments, I tell myself that the Japanese beetles are performing an important task. I am working with the Japanese beetles and trying to live in bucolic harmony with them. Usually in July, in a good season, leaf pulling and hedging may need to be done to control the growth of the vines, especially our vigorous cold-hardy varietals, and keep the fruit open to sunlight and air. The Japanese beetles destroy enough leaves that we would be dunces to pull off any healthy leaves or trim shoots. If we did so,

JAPANESE BEETLES

there would be no leaves. So we leave the leaf pulling to the beetles while we craft our attempts to control their dissemination.

Japanese beetles have brought me to my knees, my wit's end, forced me to yank on my own hair. Because they have no natural predator in our corner of the woods and meadow, we cannot encourage beneficial insects already here by planting certain cover crops or perennial plants nearby. I have resorted to bribery, enticing my own goddaughters to wander the gardens and vineyard with their soapy jars of death, paying a new, shiny copper penny for each beetle they collect and successfully drown. There is something inherently thuggish about this kind of behavior.

In the autumn of that particularly bad infestation, a neighboring CSA farmer overhead my vitriolic rant about Japanese beetles while in our local village store. He contacted me, under separate cover, at home later that evening. His suggestion for me arrived in a small package in the mail, almost like a Tupperware container full of some exotic Middle Eastern delicacy. If we could not encourage native beneficial predators, we could purchase beneficial nematodes.

Beneficial nematodes do arrive in a cardboard box in the mail, packed in ice, and must be refrigerated immediately, next to the tahini and Greek-style yogurt. They have a powdery pastelike consistency that you spoon into water to activate. I am reminded of sea monkeys, those weird little sponges shaped like sea monsters that we used to get as children, also through the mail. Just add water and they expand, to everyone's delight.

My mother swears by the nematodes. She sprayed them into the ground based on a local gardener's advice when she had been beset in her garden with the beetles. She hasn't had them for fifteen years.

Last year, with an overwhelming pack of beetles dining their way through our leaves, we sprayed the reactivated nematodes the requisite three times, trying to keep the ground moist after each application. A difficult task given that we were having a drought at the time. The jury is still out on the effectiveness of our Nematode Operation No-Beetles.

Another possible antidote to the Japanese beetle is milky spore powder. Milky spore is also a beneficial bacterium that when inoculated into the ground causes a disease that eliminates the soil larvae of the Japanese beetle. It is supposed to have long-lasting effects as well, good for up to fifteen years or so. It is usually applied in August, after the adult beetles have mated and laid eggs. The new larvae ingest the milky spore along with the roots in the soil that are part of their usual diet, and within seven to twenty-one days they expire, releasing a proliferation of more milky spore into the ground to combat any other Japanese beetle larvae.

Milky spore does not affect other beneficials or harm the soil or plants in any way. We have always heard mixed reviews of its efficacy. It is often suggested to apply it in addition to the beneficial nematodes. Depending on the success of our nematode onslaught, milky spore is something we might try in the future also.

We have one other particular unwanted visitor in the vineyard. A funny-looking, Dr. Seussian, fuzzy collection of protuberances that collect in new leaves. Lumpy little pouches that form within the structure of the leaf that color up a wild strawberry red. This is tumid gall or what used to be called tomato gall because of their tomatoey color. The name got changed to grape tumid gallmaker so as to avoid confusion about which plant these insects were plaguing.

The round little galls are the eggs laid by midges that will eventually hatch and leave the gall created in the leaf tissue. They are often mistaken for an expression of phylloxera, an aphidlike insect that may settle in the leaves rather than in the rootstock, or in both. Phylloxera has a similar-looking gall, but they are very white or the palest of mint greens. Tumid galls are generally not problematic for the vine or the fruit; they tend to be sporadic through-out the vineyard. The best way to deal with them is in a preventive fashion. Cultivate, cultivate, cultivate. It is best to hill up under the vines early in the season, in late April if possible. This can prevent the adults from reaching the soil surface. We flatten out and hill up throughout the vineyard to help keep the insects where they belong—in the soil.

There is a myriad array of insects that live in the vineyard. The major-ity of them are propitious. Ladybugs, Anagrus wasps, and lightning bugs all have roles to play within the fine-tuned ecology among the vines. In fact, in good organic viticulture, creating a habitat for beneficial insects is part of the design. To provide a habitat, plant plants that offer nectar and that provide alternative prey for the beneficials. Parasitic wasps are one of the key beneficial insects to bring to the vineyard. They thrive on nectar and prey on many of the classic damaging insects like leafhoppers and leafrollers. Cover crops are a great way to provide habitat. Planting buckwheat, alyssum, and phacelia are all good possibilities, as they have very high pollen counts and can increase the colony of wasps sixfold with the food they provide. Pollen is also an important food source for beneficial mites that eat marauding mites. Pollen is important for the honeybees, of course. Pollen, pollen, pollen. Planting perennials around the vineyard can also provide good habitat. The

Anagrus wasps even like blackberry canes and plum trees, and farmers interested in establishing a perennial polyculture often plant them.

In providing a home for "good" bugs, it's important to strive for balance. Imbalance in beneficial insects can be as problematic as an overpopulation of the ones we actively want to deter. Ladybugs provide a perfect cautionary tale. In moderation, they are effective and promote good health in the vineyard. But if they become overabundant and make it into the crush of our harvested fruit, they can destroy the wine. If they get crushed with the grapes, they release compounds into the juice called methoxipyrazines, which are believed to create offensive notes in wine. These compounds could be pheromones, released to communicate to other ladybugs. Researchers don't know exactly what they are saying; it could be anything from "I found some great food," to "Danger, danger, danger just stay away." In any case, the compounds can leave a taint, adding unsavory flavors like bell peppers and asparagus that just taste foreign to wine.

Vineyards should not be grown in a void. They are a living and breathing cultivated installation of the natural world in a particular landscape. Sadly, many vineyards today are monoculture plantings, with very little to relieve the monotony of row upon trellised row of vine. But there are those winegrowers who recognize that too much of the same thing in one spot is not beneficial for anyone or any place—nor is it beneficial for the vines or the fruit. And for making wine, in the end, monoculture ultimately doesn't serve.

In the world of wine, it is clear that certain places are meant to be under vine. Particular slopes on hillsides, or terraces leading down toward a great river, or walled vineyards—small landscapes that create a confluence of

LADYBUG IN THE VINEYARD

this thing we call terroir can create some of the most transcendent wines. Because of that history with wine, many such locations already exist. It is not to be expected that those spaces would ever be given over to other kinds of farming to create diversity. However, there are winegrowers who are creating diversity within what could otherwise quickly become a stagnant monoculture. They plant lively rows of legumes or sunflowers; they tack inspiring birdhouses up to trellis poles to encourage beneficial birds. Some, like our friends Werner and Angela, even do pull out vines here and there to create permaculture plantings within their framework, installing an apple or oak tree, surrounded by an array of perennial plants like roses, chicories, and yarrow.

In the Vine Yard of Eden

One of the most beautiful and diverse vineyards I have ever visited is in the small hill town of Dogliani in northern Italy. We were waiting for our friend Nicoletta to arrive at her house in order to walk us through her vines and taste new wines. She had been off entertaining a group of clients for a lunch and tasting. In Italy, when meeting friends, often part of the process is the wait.

NICOLETTA'S HOUSE IS semidetached, which means that the old farmhouse is divided into two homes. Her neighbor is Signore Stralla, an old farmer who has been working the land on that outcropping for more years than he'd like to count. His vineyard cascades down the hill in front of the house, Nicoletta's younger vineyard slopes down behind. At the top of Signore Stralla's vineyard is a garden, which in spring is in bloom with little bulbs and white allysum. And in the vineyard itself, apple and plum trees anchor the

SIGNORE STRALLA'S VINEYARD

ends of rows, also in white-flowered bloom. A fig tree and statuesque rose-bushes stagger down the hill. As we walked through the rows, the perfume of wildflowers and herbs swirled around us. Wild garlic, purple wild hyacinth, chamomile, dandelion, clover, and a carpet of tiny lavender flowers that I couldn't identify. There were miniature copses of wild scented geranium and mallow.

The scene was idyllic and atmospheric. An overcast day with brooding clouds. The house sits near a small church and tower, San Fereolo, set in a greensward. The bell in the tower rang three times. Birds called back and forth between trees and trellis posts. We could hear a dog barking and the voices of the two women tying the vines with long whips of reed. We stopped to chat with them. One of the women working was from Nice, where, on this particular visit, we had just been.

Nicoletta says that Signore Stralla taught her everything at the beginning. She was a city girl born to a journalist father and a ballerina mother and grew up in Milan. Her entry into the natural world came about through the acquisition of the house she divided with Signore Stralla. She didn't expect to plant vines, or produce wine. It just happened incrementally, by spending time at Signore Stralla's side in the field and at his and his wife's table, and by following the rhythms of the season that she found there on that little hill. As we sat at Nicoletta's own table talking and tasting her Riesling that speaks so eloquently of the wildflowers and herbs in this place, she admires in Signore Stralla that he is a true contadino, a farmer of the land. "All *I* want to be is a good contadina," she confides.

Signore Stralla's vineyard is entirely enchanting and, once you step in it, you don't really want to leave. And Nicoletta's vines, nestled on the other side of the hill, share in that same natural beauty. In her own farming choices, she has gone beyond Signore Stralla's teachings, and follows a biodynamic methodology. I want my vineyard to be like these, I thought to myself, alive with a natural and slightly wild life that supports the vines and makes every day standing out in the midst of the rows a pleasure. I too want to be a good contadina.

Signore Stralla's and Nicoletta's vineyards have certainly inspired much of what we do here on our own small outcropping. And I am sure that from every vineyard I have walked in—and there have been many—I have taken a piece of each of them and brought something home of their sensibility. I often steal talismans from other vineyards, sharing something of those parcels' magic with my own. From Puligny-Montrachet clos I have brought pieces of red calcareous rock from the soil; from the steep inclines of Liguria that run straight down into the sea, I have brought home the gnarled cutting of a ninety-year-old vine, now just petrified and worn like a piece of driftwood. I have brought home the pressings of the wild hyacinths growing between vines at Nicoletta's, and I have brought home a packet of poppy seeds from Austria. Then, too, I have brought home the tastes of the vineyards themselves, the curling wild garlic, the wild asparagus tucked into the edge of the woodland surrounding the vines, the spicy wild arugula picked for a luncheon salad or for a quick bite on a walk. I have certainly eaten fat figs from trees in an old abandoned vineyard and dark red rose hips off the bush in late fall at the end of the rows, as well as the fruit of the bunches themselves, slightly overripe and for some reason left behind after harvest. These memories are ever present with me in my own vines.

And while Nicoletta and Signore Stralla's vineyards are held up to me as something to emulate or re-create, there is one other vineyard whose image I hold always before me—an abandoned vineyard, the secret vineyard that Caleb and I stumbled upon while on a walk.

We were staying in one of those villages perched high above the sea on the Italian Riviera not far from the border of France. This sounds so luxurious and exotic, but it was nothing of the sort. It was November, the off-season, and we were able to arrange a good deal for a lovely apartment that would have otherwise been far out of our price range during the high season of summer. The weather was still grand: sunny and quite warm. The apartment we were renting was tucked into a narrow street at a little crossroads that went off into what looked like a wild hillside. On a mild evening, shortly after we arrived, we decided to walk down the hill and follow the road, figuring that at some point it would take us back up into the village, where we might find dinner at the little local restaurant in the small piazza across from the big church.

The road quickly took us into an overgrown palm nursery. While Puglia is considered the breadbasket of Italy, and the Loire Valley the garden of France, Liguria is the greenhouse between the two. The hillsides overlooking the water are covered in glass houses used to cultivate all manner of vegetables and flowers, and that area does a thriving business in the nursery trade all over Europe. Much of the greenery is grown inside the glass houses, but other plants that can withstand the mild climate can be cultivated outside. So, suddenly on this walk, we found ourselves in a tropical haven.

The road, relatively well trodden, petered out into a rough track. As we went forward, we spied an opening within the palms, mostly hidden by their fans. We saw a garden tucked within: an overgrown garden with three or

four long rows of artichokes, some squashes rotting on the ground, a covered hoophouse with the door left wide open and the inside in disarray, and a small vineyard with fruit drying on the vine. It was as if the owner had suddenly been called away, left in the middle of hoeing a row of carrots, fully expecting to return. But no one had returned.

The garden had not been derelict for too long—maybe a month or so. And we wondered at the disappearance of its farmer. Perhaps a sudden illness, or a death. A story lurked. Those dark grapes hanging, some still good enough to eat, were melancholy in their forgotten ripeness. We picked several bunches to roast the next day with sausages and onions.

We stood for a long time in that garden, a perfect paradigm of a diverse agriculture. Even with its sad air, it was comforting to be nestled in this space surrounded by the palmettos. It spoke of thought and hard work, the rewards of bounty. The farmer would have enough here to feed the family for the year, plus a surplus to sell, in addition to making the home wine to be stored in the cellar until the next harvest. We wondered what variety the grapes might be. In Liguria, there are over a hundred obscure indigenous vines that can be found exactly in situations like this. A farmer has five or ten rows of something that no longer grows anywhere else in the world, something almost extinct but still surviving, still tended. The thought that these grapes might be something undiscovered, or nearly forgotten, was exciting.

We didn't want to leave. The desire to pick up the turning fork, to clean up the squash bed, to harvest the rest of the grapes, to tidy up before leaving was strong. We imagined what it might be like to stay here, to rent the apartment long-term, to find out the narrative behind this garden, to take up its task. The fantasy was very seductive. And, at another time, it might have been possible. It might have been that split in the road taken. But we already had a hoophouse, and we had planted our own small orchard and more than ten rows of vines. We couldn't imagine abandoning our own clearing in the meadow, leaving it in dejection. Instead, we took the memory with us, and it has become part of the assemblage of our own layers of garden, of physical planting, of physical labor, and the narrative of the other gardens and vineyards that inform the shape and intention of our efforts.

Orchard

June has arrived, and we walk down the narrow rows of our vines in the vineyard. The vines are flowering and the field is scented with the white pepper aromas of the blooming La Crescent and the gardenia-like perfume of the Marquette. We cross one of three bridges. Two were built from that seemingly never-ending supply of old cedar from that original garage. The third is an earthen bridge covering a culvert.

THE EARTH AND SOD WERE MOVED from the ground where we planted three trellised panels of young apple whips at the end of the rose garden, creating a wall of blossom and fruit. The bridges cross a small waterway that we built with a swale that collects water from the road and land and funnels it down to the lower quarters of the meadow, eventually feeding back into the seasonal brook that defines one side of our parcel. We have begun to plant

grasses and reeds in the floor of the ditch; any rain and runoff from the road can course through the vegetation, which helps clean the water.

But you could come to visit just as easily on a spring day, and we'd find ourselves in the cherry orchard, the small trees laden with fragrant white flowers with white and yellow centers. Beyond the cherry trees are those espaliered apple whips dividing the cherries from the rose garden. To the right of the rose garden a line of younger apple trees marches toward the apple orchard. Through the fretwork of their branches we can see the winter greenhouse, or hoophouse.

The apple orchard surrounds us. A handful of pears, plums, and ornamental crabs are planted here and there for variation and diversity. The mature orchard is defined by two old trees that were present near the house when we bought it. We call them *Les Amants*, The Lovers. The two wild apple trees twine around each other in a tangolike embrace, their tops forming a perfect broad canopy. They bear the first apples of the season, dropping to the ground before the end of the summer. The fruit is the color of August, a golden yellow, making a sweet, pulpy juice that tastes exotic. They are best for cider, really too sharp and acidic for eating or cooking. They are not keeping apples; they don't last long in storage. Maybe it's the heat of the end of summer that leads them to rot so quickly, but they are an ephemeral variety. We must process them quickly to capture their silky and textured flavors. They always come as somewhat of a surprise. When the apples begin to fall, we are never expecting it, even though their drop happens at roughly the same time each year. We are still in the thick of the growing season, busy tending the gardens and vineyards. We are not quite ready for harvest or the trappings of fruition. But somehow we manage to collect the small golden fruit during the height of the summer.

This summer we had a little help from a young buck living in the meadow below the house. Yearling males that have grown enough in maturity and have begun to develop their horns are usually pushed out of their mother herd to begin to fend for themselves and round up their own herd. This yearling seemed to be newly independent, making two daily visits into the orchard. He would walk calmly around the lower walled vegetable gardens and straight to the early-dropping fruit of *Les Amants*. There, he would delight in breakfast or dinner. He was very tame, willing to share quietly the orchard space around the house with us. After his supper of apples, he always meandered through the rose garden and out through the vineyard, giving us good long looks at his beauty and grace. He never bothered the vegetables or the vines, only ate grasses and the apples. We figured as long as he kept to those two delicacies and left enough apples for us, we were happy to share in the bounty of these two trees.

Original to the property are several other old orchard apple trees—wild pippins (ones that grew from seeds or "pips") that are beautiful cider apples, small and hard, with a pure yellow skin or striped and blushed with shades of red. Some drop fruit early in the season like *Les Amants*, while others drop late, and we tailor our picking and pressing to the rhythms of these old grande dames.

At the bottom of the meadow is our oldest tree; she anchors the point at which the meadow reverts to stone and forest. She is a large tree, maybe a hundred or so years old, and has clearly seen less complicated days, when the forest was open pasture for sheep and the younger version of herself had space to stretch in the sun and wind, only girdled by a low stone wall. Now she is hemmed in, copper beeches and maples crowding her. Early on in our tenure at the farm, we did some remedial work on the trees along a stone wall that divides a part of the lower meadow. With saws in hand, Caleb and a friend cleaned up along the wall ending at the old apple, but even still, she needs more work. Half the tree is thriving with multiple new shoots reaching for the sun. The other half is gnarled and blighted with algae and a strange variety of moss that looks like a small, wild orchid, a plant sustained only by air.

She blossoms up high in her canopy, and a couple of years ago we got a fantastic harvest that dropped obligingly to the ground. Last year, the weather was such that we had no apple harvest from any of the trees, and this year, though we've had a good harvest, the old apple produced minimally. It's time to prune away the dead weight and clear some of the small trees choking out the sunlight and elbow room. This will be late autumn and winter work, when the trees have gone dormant and the demands of the other plants are less.

When we planted our cultivated orchard that wraps around the house, we knew we wanted to use the apples for the kitchen at the restaurant, but we didn't know then we would fall in love with traditionally styled sparkling cider and want to make it ourselves. Initially, we planted primarily Liberty apple trees, a cross between Macoun and a variety unsentimentally designated Purdue 54-12. These trees are disease-resistant and produce fruit good for both cooking and, luckily for us, cider. Macoun itself is a cross between McIntosh and Jersey Black. And Jersey Black is an old heirloom North American variety that is now quite rare but was once used for cider. They were identifiable by their dark red, almost black skins. Purdue apples come from

a breeding program that started in southern Indiana in the 1940s. Liberty seemed a fitting choice, since one half of us (Caleb) is from Vermont and the other half (me) hails from southern Indiana. And, although we didn't realize it at the outset, Liberty apples have come to be a lovely way to express and blend our own personal landscapes.

The orchard came to the property before the vineyard. For a long time we had held an image of a house in an apple orchard as a place we wanted to live. For me, it's the same kind of evocative scene as the image of sunken stone gardens made from old, abandoned cellar holes. It stays with me, creating an atmosphere, a feeling, a sense of space and landscape.

When we were first married, living in another town, with a slightly different life, friends had told us about a farmhouse that was for sale down the road from them. When we drove down their road to investigate, we saw a small but elegantly proportioned Cape-style house ensconced in an orchard.

We weren't really looking for a house then, but something about an old, derelict farm in need of fixing up began to pepper the domestic conversations between ourselves and our conversations with friends while sitting around the dinner table. Caleb had grown up in an old farmhouse that needed a lot of attention, but for my part there was no real sense of what it meant to buy an old place. However, I had firmly embedded the notion in my mind, much like the orchard house was embedded between the apple trees.

Our timing was all wrong for that little house in the orchard, but our knowledge that it existed informed so many of our future decisions. We didn't

set out looking for a house in an orchard, but the idea was there, always present when we finally were ready to find a spot of our own.

When we arrived at our house for the first time, before it was ours, it fulfilled none of our preconceived or imagined hopes. The house was not old, but instead had been built in the early 1970s, and it seemed very shy on any kind of charm. The land had not been worked in years, and in fact had been left to redefine itself eventually as forest. At first glance, there was no orchard, there were no garden beds, there was no barn, there were no climbing roses, there was no vineyard. But the land was broad and open, with a 70-mile view to the north and east. Layers of hills drew the eye to the snow-cap of Mount Lafayette in New Hampshire, out there on the horizon.

The house sat awkwardly on the property. At the time, the stone wall dividing the lower meadow reached all the way up to the road, separating the house from its own land with stones among a wild, messy thicket of trees and bracken and a forgotten hedge of heirloom roses. The open fields to the north did not belong to the house and, in fact, had always belonged to the neighboring farm.

In truth, I don't really know why we decided on this piece of land and small house that looked more trouble than it was worth. The long view clearly called to us. A friend of ours who was also in the market for a new

home had looked at it around the same time, and she revealed to us long after we bought it that she had thought the project so daunting that when she had driven into the driveway to meet the Realtor, she paused and looked, then promptly backed out onto the road and drove away.

Like most things in our life, the prospect of this place seemed to speak to us in a way that drowned out the voices of perhaps better, more steadied judgment. Somehow, through the lens of the dark wood Quebecois log-kit home and the wilds of the meadows, we must have seen an orchard and some roses rambling up the corner of a cottage. The ghost of possibility must have whispered to us and beckoned powerfully. At the time, I felt almost like I was in a weird and witchy dream state in which we and the house and the land existed in a slightly alternate reality, and our task was to uncover what lay beneath its reflection. Even though it appeared far from enchanting, the house and land somehow cast a spell over us. Before we knew it, we were the proud owners of a summer camp barely converted to winter living, white-washing the floor and walls in the month of January before we moved in.

The orchard came quite a few years later. There were so many things to be revealed before we could see the embrace of the apple trees. But see the embrace we did, and we began finally planting the core orchard of twelve Liberty trees, and then kept planting. Now there are fifty-six trees, most

mature enough to be heavy with fragrant blossoms in spring and bearing fruit in autumn.

While there are the workhorse Libertys, and the elegant plums and cherries, we have also planted more pear, plum, and apple trees with a focus on heirloom cider varieties that respond well to our landscape: apples like Hidden Rose, Cox's Orange Pippin, Old Pearmain, Frequin Rouge, Wickson, Reine de Reinettes, and Orleans; perry pears like Brandy, Barland, and Normannischen Ciderbirne; and plums like Superior and Toka. The names alone provide succor and poetry. It is one of the things I love most about working with plants, the common vernacular names that conjure up a landscape of Arcadian bounty.

The apple is a magnificent tree. *Malus domestica* is the botanical name for this member of the rose family, or Rosaceae. The tree originated in Central Asia, in the Tien Shan mountain forest region between southern China and Kazakhstan, where its wild ancestor, *Malus sieversii*, still exists. The center of diversity for the genus *Malus*, though, is in eastern Turkey, where it is believed the apple was first cultivated. It is thought that Alexander the Great brought back apples from Kazakhstan. Winter apples, key to our northeastern landscape, have long been important food staples in the colder climates of China and Europe.

Domestic apples came to us here in North America on ships bearing the early colonists in the seventeenth century. The first orchard on record was planted in Boston by Reverend William Blaxton in 1625. The only apples native to North America are crabapples, the small, bitter wild fruits that make a rather haunting cider.

Most cultivated apple trees in an orchard are propagated like grape vines, by taking cuttings from prunings. The cutting duplicates the mother plant. Wild trees that grow along stone walls or that stand resolutely in the center of old sheep meadows may have come from once cultivated orchards, or from crabs, but they also occur naturally. The tricky and smart thing about apples is their five seeds. Each seed holds a deep well of DNA, genetic information that can be traced back all the way back through the centuries to the Tien Shan forest and its 60-foot-high wild apple trees. When that seed gets planted by birds or other animals that recycle seeds through their digestion, like our little buck, the plant that grows from that seed reaches down into the blue waters of DNA and adapts the tree to the particular terroir it finds itself in. Certain attributes will be highlighted for its survival, others will recede, each seed creating a new and unique variety, no longer resembling the mother tree. This is the nature of adaptation and evolution.

Most of our apples, both the cultivated and the wild types, are used for cider and cooking, with the majority going into our cider. Cider is going through somewhat of a renaissance these days. Local orchards are growing and pressing cider from a wealth of new and old varieties, and finding new ways to survive and thrive. New cideries manned by young farmers and fermenters are bringing back the old traditions of cidermaking and drinking. The uncovering of the history of cider in our culinary narrative is both exciting and inspiring.

We came to cidermaking after we were struck by the notion of winegrowing, even though we had the orchard long before the vines. The same year we planted the first block of vines, we came across a notice for an autumn cider tasting at a local orchard and cidery. Curious, we decided to attend. For some unremembered reason, we ended up being unable to go on the appointed date, so we scheduled a private tasting and met the owners.

The tasting was structured just like a wine tasting at a winery. We stood in the cellar sampling new pressings from that fall as well as already bottled cuvées. There were single varietals and blends. There were ciders that had fermented completely in stainless steel, and special reserves that had spent some time fermenting or aging in barrels. The orchardist specialized in bringing back and cultivating heirloom varieties. There was a range of interesting and complex tastes. They were in a fresh style and released within that first year of harvest, the producer believing that cider is best drunk young and not really meant to age for too long, though he has one cider that he leaves to age in a barrel.

This seemed a brilliant idea to us. While we waited for our grape vines to produce, we had a ready crop of other fruit to ferment. We could make cider, get it out into the world, start learning and sharing.

Most ciders we tasted in our initial trials and travels were bright, refreshing, and young. They were tasty, but very different from the experience of tasting wine. They did not necessarily have the same depth and intrigue.

We ended up learning to make cider from an older Vermont farmer who lives about a half hour from us. He is the grandfather of a friend of ours, a terrific cook and chocolatier named Erlé in honor of his French-Canadian ancestry, and when he heard we were interested in making cider, he reminded us that he and his family had been making barrel-aged apple libations for generations. He came to the osteria one night to sit at the bar for dinner, and brought a bottle (an old Goslings Black rum number that he'd co-opted from the recycling bin) drawn off his own barrel for us to try. There is a long farm tradition in Vermont to bottle both farm-fermented cider and maple syrup in finished liquor and large-sized beer bottles. There's something slightly and beautifully rebellious in this.

We were not fully prepared for the still cider Erlé brought us. We had tasted nothing like it; it was stunning. Rich and nuanced with a fluid texture and a heart we had yet to taste in cider. I wanted to know how to make cider like that. So we went to go see Erlé's grandfather Kermit.

Kermit has been farming all his life, a farm that had been passed down for a number of generations. He had retired recently, and we went to visit on a cold, dreary March day. We tasted from various barrels, all old whiskey or bourbon barrels brought up from Kentucky, except for two barrels that had once cured ginger. Those barrels had been an unintended addition, but ended up making an intensely baroque ginger cider.

Kermit's family makes cider from only wild apples gleaned from the side of the road, out in meadows, deep in the woods. He taught me that the best cider is made from these incredibly varied trees, and each year the cider produced is a blend with flavors typical to only that vintage.

Kermit's philosophy is not unlike the many winegrowers I have met and learned from over the years. Less is more. Little to no intervention in the process. They make their cider at the farm, by picking, pressing, and fermenting in their wooden barrels they've collected over years. Kermit is adamant that cider needs time in the barrel to coalesce and mature. He told me that cider gets a little interesting after three years in barrel, and doesn't start getting good until year six.

At the end of that first visit we sampled an infused cider aged almost sixty years, a brilliant amalgam of unidentified flavors and textures that were still bright, yet with a caramelized acidity—a gestalt of whatever went into the blend of apples, herbs, and flowers back in 1956. No oxidation here. These were bottles of a cider that had been saved from the cellar of a neighbor's house that had burned down. In Vermont farmhouse tradition, the bottle's original label was for a Three Monks Trappist beer.

Tasting cider with Kermit changed everything. So much for the idea of having an easy, early cider ready to go to market within the first year of pressing. I wanted to capture what Kermit and his family made in the cellar of their farmhouses. The challenge would be to figure out how to do so in our own cellar.

Another man too has greatly informed our philosophies and efforts at cidergrowing. I found Eric Bordelet through the winemaker Randall Grahm. Randall has been a very supportive friend to the farming of our wine here in Vermont, and he himself now produces a cider made from apples, pear, and quince. Randall had told me that if I wanted to make cider, I needed to go to the source. He hailed Eric's *poiré* (pear cider or perry) as "the most sublime fermented beverage on this planet," and Randall himself is trying to create something of the same ilk in his own cellar.

Eric lives in Normandy. A talented and fêted sommelier in Paris in the 1980s, he returned to his family farm at the end of that decade and decided to make cider in the same way great winegrowers make wine. He has done just that.

The family farm is tucked into a fold of rolling green hills and is set rather atmospherically amid the ruins of a great Norman castle. The castle is too far gone to restore, but Eric does use the cellar beneath the tumble of walls to age his ciders, and he makes some truly melodious ciders from old trees. His poiré is a blend of twenty-five different heirloom pear varieties, and some of these pear trees are over three hundred years in age. It is from him, and the layered flavors of his ciders, that I learned that true cider is an amalgamation of different apples or pears that bring different elements to the mix. No one apple or pear can make a perfectly balanced cider on its own. Cider is not like a single varietal wine. Certainly some very nice cider has been made this way, but on Eric's palate, and now on mine too, ciders made from a blend of varieties that hit notes of bitter, sweet, salty, acidic, and tannic create a more symphonic collage of sensations.

Eric has a good orchard of these old trees, and also plants new ones in order to grow and keep the orchard moving forward. He works biodynamically, in every sense of the word. The orchard floor is littered with delicate wildflowers: scabiosa, cosmos, poppies. Our friendship was sealed when we asked if we could see his compost.

Eric works in many of the same ways as a biodynamic winegrower. He works from the soil upward. The new trees are all trellised; the older ones were planted in a diamond-shaped pattern. He focuses on making biodynamic compost; he uses plant medicines when necessary.

In the cellar, he works minimally, having figured out what fermentation program works best for his varieties in his climate. He has been greatly influenced by the winegrowers of the neighboring Loire Valley. The yeasts are wild and the fermentation occurs in a tank and is bottled directly, at just the right moment to capture a soft mousse and a pearlescent bead. He shares information, but not too much. He believes that every cidergrower should develop his or her own process, responding to the needs of the cider, and not try to emulate other growers or try to follow a set program. For him, this is the craft of cidergrowing: the grower working with the trees and land, and creating the best environment in the cellar.

Land, Soil

The land that is best for an orchard is one that mimics a forest-edge ecology. We are fortunate that our farm is at the meeting of field and forest, for this provides a good and diverse environment and also a ringside view of what happens when forest meets field. Much like the discussion about soil for vines, organic and biodynamic growing in the orchard relies on healthy soil to manage and maintain the plants. Forest-edge ecology has a specific set of circumstances that creates such a fruitful environment for fruit trees.

ALL OF OUR WILD PIPPINS occur on the edges of forest or the edges of tree lines, as our field is encircled by trees, and there is no mistake or randomness in this occurrence. The evolution of soil follows a specific pattern; we can see it as we watch newly tilled soil shift and change. Beneficial leafy plants arrive first—the pigweeds and chickweeds, the dandelions and chicories—feeding the ground with decomposed organic matter at the end of the autumn. Those leafy plants give way to more woody characters like blackberry, wild rose, and trees like poplar that grow quickly and profusely. If that tilled patch were allowed to continue developing toward forest, those little trees would grow taller and heartier, blocking the sun and therefore blocking the growth of the leafy greens, creating a floor defined by low-growing shade plants and leaf mold. At the edge of that forest, where still a bit of sun filters, the wild apple will grow.

What's happening to the soil that makes it so special for apple trees? The presence of all kinds of organisms working hand in hand makes the apple tree's world go round, especially those organisms classified as fungi.

In winegrowing, *fungus* is a bit of a bad word, and certainly when it lives on your foliage or your fruit, things have clearly gotten out of balance. But many fungi that belong in the soil aid and abet the delivery of food and mineral nutrients to plants. In the vineyard, we talk long about mycorrhizal fungi, and this type is crucial in the orchard too. It's crucial in every growing situation. Its skeins of hairs traveling through the soil provide that synergistic roadway for food and also act as a shield below the surface, creating a natural antibiotic armor between the pathogenic organisms in the soil and the plants. In favorable weather conditions, these fungi send up flowering bodies, which we recognize as mushrooms. On the edges of our field is where we find edible mushrooms when the weather creates a perfect storm for the fruition of the fungal environment.

In an ideal soil scenario, the fungi would outweigh the bacteria by about ten times. And the majority of these fungi are the beneficial ones. Pathogenic fungi do also exist in the soil, and sometimes they are more helpful than detrimental, even though farmers and gardeners of all stripes tend to demonize them. We will never rid ourselves of the pathogens and, as I've written before, we shouldn't even try. They have a role to play in the intricate soil web that creates plant life. Just because they *can* cause havoc and feed off the roots, leaves, and fruit of plants, doesn't mean we should exterminate them, or that they *will* cause havoc. It is when conditions create a soil environment that is out of sync or alignment that pathogens are able to prosper out of their normal relationships.

The difficulties for the grower in this day and age are that the conditions are constantly evolving, and evolving very quickly. Sometimes the soil and the plants don't have time to catch up and figure out how to manage themselves in these new circumstances. Our own northeastern climate is going through all kinds of change. In Vermont, we've tracked the growing seasons in the last ten years and found an untrustworthy trend toward heavily wet springs and first half of summers, then an abrupt shift to droughtlike conditions. In this kind of dynamic, thank goodness for the extreme dry weather. It allows us to catch a break in bringing the soil back into balance: Heavy

water recedes, the temperature and dryness don't allow the pathogens to proliferate, fruits and vegetables get sun and air; they can successfully ripen, drawing on a backlog of water if need be. It seems to be a time of extremes in the world above the soil as well as in the world below.

The further difficulty is that now, after ten years at our home farm, we might feel comfortable in the knowledge that we can see a pattern, and respond accordingly and pre-emptively, but there is little consistency in the global weather, other than knowing there are these extremes. What we might think has become a seasonal shift in the kind of weather we have, something we are stuck with for a while, may change next year to something completely different. It is doubly important to be able to respond immediately to the given circumstances. The grower must be fully attuned to what is happening day-to-day. The grower must be able to look at the small scale of that season, that month, that day, as well as a few years ahead to the future health and longevity of any perennial crops. It is not an easy task to need to be everywhere at once.

Soil in the forest and on the edge of the forest gets where it wants to go by a simple process: The leaves and limbs fall and decay, decomposing into a heady brew that in turn feeds the soil, then the roots of the tree and the new leaves and fruit (or nuts, or seeds) the following year.

Preparation of your land for an orchard usually happens a couple of different ways, just like in the vineyard. The yard or field gets plowed and tilled up, the trees get planted in the opened ground, and growing and caring for them begins. Or, you plan ahead and give yourself at least a year to prepare the ground with plowing, tilling, and cover cropping. The same cover cropping that is done to prepare a vineyard site also works for the orchard. Cover crops give your orchard a leg up on building organic matter in the soil, correcting any soil imbalances, and feeding a fungi-dominant humus. Clovers are always a good cover crop, and the dark clovers, like red or purple, tend to offer more for mycorrhizal fungi than white clovers. A seeding of buckwheat not only helps develop nitrogen in the soil, but is also good for cutting back on those runner roots of native grasses that like to create mats of congested growth.

There is a preferred schedule to this process of preparing the soil. Tilling in August and seeding clovers and or buckwheat helps a first growth of cover crop before winter comes. By late fall the cover crop can be cut down. (Old-time orchardists like to scythe.) Leave the cover crop on the ground. Soil and trees like decomposition, and the vegetal matter left over the winter will work into the soil, supplying nutrients in the spring. The roots of the cover crop will still be in the soil, and come the warmer months, there will be a second growth.

In the spring, the areas for planting the fruit trees can be forked and dug up. The soil will already be loosened and crumbly. A further seeding of herbs and plants that have healthy taproots will also help improve and break up the soil naturally.

Michael Phillips, a holistic orchardist based in northern New Hampshire, has written thoroughly about another composting method that is extremely beneficial in the orchard: ramial wood chips. A woman named Celine Caron in Quebec was instrumental in guiding the research on ramial wood chip compost work at Laval University. Her research and observations ended up in the book *Ecological Fruit Production in the North,* and both she and her book heavily influenced Michael Phillips's own work and writings.

Their philosophy posits that a diet of ramial wood chips decomposing into the soil around orchard trees becomes the mainstay for feeding beneficial fungi, which in turn support the orchard trees. The tops of deciduous trees and woody bushes are pruned, then run through a chipper, and the nutrients present in the green cambium and bud tissues of these new growths include a delectable cocktail of nitrogen, phosphorus, potassium, calcium, and magnesium. When broken down into the soil, these provide a virtual smorgasbord to the soil fungi, which in turn feed nutrients back to the roots of orchard trees.

In the orchard, this kind of mulch is preferable to composted manures or vegetation. We're going for forest-edge ecology for the fruit trees, as opposed to providing an animal-based support through manure for the vineyard. And for the home orchardist, there are a plethora of mulches available on the market that you might think are just as good to use as ramial wood chips. However, there is an important distinction to be made here in terms of providing the right kind of food for the soil. The ramial wood chips from the tops of deciduous trees and bushes are preferable to the pine barks used for landscaping beds that can be shoveled up into your pickup truck or the back of your station wagon. Also, give the bagged cedar mulches you can buy at the nursery a wide berth. Ramial wood chips are subject to what is known as white rot, which in turn supports deciduous trees. Brown rot, which follows from pine and cedar mulches, is highly tannic and is the foodstuff of the evergreen forest—not appropriate for the deciduous fruit trees. Red-colored mulch from the nursery is also to be avoided. That hue comes from a chemical colorant that then leaches into your soil, adding an unwanted chemical reaction among all that good bacteria and fungi.

Another kind of mulch to be wary of is black plastic or landscaping fabric. It's the same in the vineyard. A wet, bacterial environment grows underneath the fabric, shifting the balance of the soil environment away from fungal toward bacterial. Any ramial wood chips or compost that you might put on top of the plastic or fabric can't reach the soil. Then the soil underneath the fabric compacts, as there is no hand-tilling, and no long taprooted plants growing to break up the soil. It makes it very difficult, if not impossible, to create a healthy soil for your orchard trees.

Ramial wood chip mulch is not the only mulch that's good to use in the orchard. We like to also incorporate straw or hay in the mix. Most typically we get hay, as that's what the farmers around us grow for their horses and cattle. Or we rake the mown hay and grasses from our field or the neighbor's and use that. It's important to understand too that fresh ramial wood chip mulch and fresh hay operate slightly differently than piles that have been sitting around for a while and have aged. They are all good, but they will provide different advantages.

Cutting-edge orchard keeping calls for mulching around your trees in what looks like a rather indiscriminate pattern but in reality is a considered mix of the ramial wood chips and straw, applying these mulches in the autumn. If you look at the space around the base of the tree as a square, on one side you might apply anywhere from 2 to 8 inches of ramial wood chips and on the other side, directly opposite, a healthy padding of straw. Next year, you would rotate the two mulches, and the third year you'd hit the alternate sides, and rotate them in the fourth year, so you would be hitting all four sides in a four-year cycle.

This kind of thinking provides the plants with a varied diet and keeps the soil balanced around the tree rather than heavy in certain nutrients on one side. What we're trying to do here is emulate nature's ways and provide the most natural environment that we can for our fruit trees.

The land for an orchard largely follows the same strictures as it does for a vineyard. You look for well-drained soil, sun, and air circulation. Fruit tree roots need room to stretch out, and their budding and fruiting will do better if they get full sun and are not too shaded by taller trees surrounding them and blocking out their light. You put in a hedge or a wall to block excessive wind. Pear trees can withstand a mild case of wet feet, whereas apples cannot. Sandy, dry soil or wet clay soils can both be improved significantly by mulches and natural sprays. Excessively wet ground may need to be addressed with drainage or plantations of water-loving plants nearby to help drink up the water.

Orchards, like vineyards, are better off on a slight slope, but can function perfectly well in a flat field. Our orchard is on a very languid slope, and it has helped to address some of the drainage issues we have on our land because of tile drains buried underneath the soil between the orchard and the house on both sides. Not everyone has the most ideal location to plant an orchard, a vineyard, or a garden, but we work with what we've got, and if we keep in mind basic, biological common sense and permaculture concepts, we can go a long way to sustaining our own larders or making a livelihood.

Varieties

O rchards are in so many ways similar to a vineyard. Their resemblance is something that I think makes them good companions in terms of farm diversity. You don't have to learn a whole new language or set of rules to grow either of them well. When seeking which varieties of tree fruit to plant, the same questions get asked about apples as they do for grapes.

CLIMATE IS ONE OF THE FIRST ELEMENTS to consider. In Vermont, we live in a generally alpine climate. We know we will have cold winters, and in the past we've been known for snow. We hope that we will continue to have snow, as the snow acts as an insulator, an important factor in our northern and native plant health.

And just like any region, we also have microclimates specific to whether you are in the southern part of the state, the northern, or near Lake Champlain. Our climate in eastern, southern, and central Vermont is very different from that in the Champlain Valley. We work two vineyards in the valley as well as our home vineyard on Mount Hunger. At the home farm, we harvest grapes a month or more later than we do in Vergennes near the lake. Our home farm blossoming time also happens about three weeks later. Then you have to factor in high altitude or valley living. Low valleys tend to get frost both later in the spring and earlier in the fall. The village of Woodstock, which is home to our osteria, always gets frost in the autumn before we do twenty minutes away and 500 feet higher in Barnard. We can start working our soil and setting out plants a little before the gardens down in Woodstock as well. Then, on top of these generalities, you have even more distinctions.

I once lived for a summer with my family at a place outside of Woodstock that was documented as one of the coldest spots in the state. We would have days of rain and entirely different weather patterns in the summer than the village experienced just ten minutes down the hill. It is rather awe inspiring how much difference a few miles and a few feet of altitude can make.

While thinking about climate, the maturation time of the fruit should be taken into consideration. Blossoming times correlate directly with climate fluctuations. In an area that tends toward later spring frosts, making sure the varieties planted don't blossom early will go a long way to minimizing the risk of damage.

The same goes for harvest time. Varieties that can ripen in the number of days a season holds is crucial. A mild autumn frost can actually be beneficial for apple trees, the cold deepening the color and also concentrating the sugars in the fruit, but too much cold and too many frosts can turn the apples toward vinegar before you even get them off the tree. Choosing varieties that have the right number of growing days for a zone is a must for a happy orchard.

Many fruit trees need a cold hibernation in the winter in order to fully go to sleep and wake up ready for spring. They are not unlike bears, requiring several months to slow down and rest so that they are ready to burst forth with renewed energy when the weather changes. Buds on branches actually require a certain number of cold days in order to break their dormancy come spring. This little built-in feature helps the trees to not wake from their deep sleep during a warm day in midwinter. The same feature requires that the tree experience sufficiently cold temperatures in order to be released from its inactivity. If the temperatures don't get cold enough, the trees can suffer. Bloom times may be delayed or extended, the leafing may be drawn out, the fruit-set reduced.

In these uncertain times of climate change, the fluctuations in temperatures and usual patterns can wreak havoc with the normal flow of the season and a plant's physiology.

Old heirloom varieties do very well in our northern clime. They are suited to the progression of our seasons and naturally resistant to the disease challenges that are currently present. Planting a fruit tree that's not resistant to our host of pests will become a bane to the gardener. Many newer varieties have been crossbred to select the prime genes to protect the trees and fruit from the worst of the pathogens we can experience. And while it's good to keep these varietal strengths in mind, a rainy season will make demands on a plant that even inbred disease resistance will have trouble fighting, given the constantly changing world of fungi and bacteria that seem to find new ways to adapt and infect.

We know that fruits grown from seed will create a whole new cultivar, but to propagate a proven cultivar, a cutting must be taken. Typically, cuttings are not own-rooted, meaning that the mother scion is not encouraged to take root, but rather they are grafted onto another variety that is known for rooting well. The main reason for this is that good fruiting trees tend to be inconsistent in taking root straight from a cutting. There are trees specifically grown to be rootstocks for grafting with scions of the desired variety, and they can help a grafted tree take hold in a landscape and confer a varying degree of vigor and disease resistance. Dwarf trees, or varieties grafted onto dwarfed rootstock, tend to be very appealing to the home gardener with limited space, but they can be a lot more labor-intensive too, often needing staking, watering, and more attention than what standard (full-sized) or semidwarf or even seedling trees require. I think in the end it may be better to manage your tree vigor with good pruning rather than a truly dwarfing rootstock.

Grafting is an art, and whole books have been written on it. I won't delve into the process here, as grafting is something we haven't come to as of yet, though we love the idea of propagating our own wild trees. There are several excellent resources already out there to guide you. Michael Phillips's book *The Holistic Orchard* is a terrific, easy-to-read companion. Another of my favorites is *The Grafter's Handbook*, by R. J. Garner, originally published in 1946 and recently re-released with updates by orchardist Steve Bradley.

In the vineyard, due to the phylloxera problem, vines in Europe are primarily grafted, and even here in the United States many *vinifera* vines are grafted, but more and more winegrowers are planting own-root vines, believing that this creates the strongest plant. In the orchard, the reasons compelling one to plant own-root are not enough given the low success rate of the trees. You find more orchardists planting by seed than by own-root. Since I have not attempted the grafter's craft myself, but I have been rooting grape vines, I am intrigued by the idea of trying to create some new apple trees by grafting the old and wild varieties that pepper our tree lines and orchard onto new rootstocks. I figure the worst thing that can happen is it won't work, but maybe, just maybe, some of our grafted trees will flourish. Growing seedling trees is in the plan as well, but that takes time, and I struggle with patience.

I love the varietal names of fruit trees. I will admit here that some of those we've chosen to plant have seduced us by the melody of their names just as much as their hardiness and their affinity for cider. Some people buy wines because they like the look of the labels. I plant fruit trees because I like the sound of their vernacular names. We shouldn't negate the allure of language. In a funny way, I think the sensory experience of growing trees, vines, and other plants also extends to how what we call them sounds. Common plant names are evocative. They can conjure and inspire, and are an often over-looked element of our landscape. How can you not be enticed by apple trees named Brown Snout or Wickson Crab, or a pear called Rousselet de Reims?

I realize this may also be one of the factors in why I chose to specialize in more obscure native Italian varietal wines as a sommelier. I love the sound of Muscato d'Amborgo, Lacrima di Moro, Pignatello, Coda di Volpe. When the wine is good, there is quite a bit of pleasure in calling it by name. I think it is the same for cider.

HUGELKULTUR

In Germany, Austria, and eastern Europe, permaculture farmers and orchardists use a technique called *Hugelkultur*, where planting mounds, or raised beds, are created, filled with decomposing wood and other vegetation. Austrian permaculture icon Sepp Holzer has been doing it for decades. Hugelkultur, roughly translated, means "mound culture." While it has become all the rage as the best new thing, it's believed to have roots that go back to an ancient form of sheet mulching in eastern Europe. The wood-filled mounds supply each orchard tree or garden plantation with a strong brew of organic material, nutrients, and air pockets for the roots. As time passes, the raised bed decomposes into an incredibly rich soil, loaded with activity. As the wood shrinks over time, it makes more air pockets, creating a natural tillage not unlike the dead taproots of dandelions. In the first few years, the slowly decomposing wood will create some heat and can extend the growing season. The woody matter helps to keep the nutrients from leaching into the groundwater, and it helps to hold water, just like a bark mulch will. Many forward-thinking gardeners believe that this is the solution to growing fruits and vegetables in dry climates like the desert where there is a need for irrigation or in colder climates that have shorter growing seasons. This especially works well if a ready source of fallen or harvested trees from which to draw exists.

There are a couple of things to keep in mind: Again, as with ramial wood chips, be careful of what kind of wood gets used. Cedar and pines will be too tannic; black walnut and some other trees can release toxic chemicals into the soil to inhibit the growth of nearby competitive vegetation. Trees like alder, apple, cottonwood, poplar, and birch are all good types.

When designing an orchard— or garden space, since hugelkultur can be used for any kind of growing—these raised beds can be made quite high and peaked or can be lower, rounder, and flatter, more like the height of a traditional raised bed we might see in a French potager. Others are built into a trench, so that they have a lower profile.

It's extremely important to understand the process and that other compost added to the wood is key to this system working. The bulk of the mound is wood, but as the wood first starts to deteriorate, it will take nitrogen from the soil rather than deliver it, and without additional nutrients the fruit trees or plants would starve in the first year or so. It's key to have a good mix of already prepared compost and vegetation that will give the plants the nitrogen that they need while waiting for the wood to decompose to the stage where the decomposition starts to give back nitrogen to the soil. The technique also takes into consideration that as the plants grown in a hugelkultur defoliate or die at the end of the season, that green material will feed the nitrogen-giving compost.

Sepp Holzer, who is growing amazing things in a colder, northern alpine zone, has implemented hugelkultur with some brilliant flourishes, encouraging a microclimate in each terraced garden with high plateaus and swales between. Depending on the situation, he might line the lower slope of the inward-facing swale of the mound with stone to capture more heat in the summer, which will radiate back into the tree's terrace. On the high banks of the mounds on the outer side, he has also planted lower-growing fruits beneath the fruit trees, which benefit from the mix of organic matter on that side. Plus, the companion plants create a beneficial atmosphere.

Along these same lines, orchardist Michael Phillips has gone another step further by creating his experi-

mental hugelkultur with buried burnt woodsy debris. He makes a crude *biochar*. Biochar, or blackened carbon (from trees, bones, you name it), is often employed in cutting-edge organic farming to provide long-term fertility. This carbon provides a haven for mycorrhizal fungi and various related bacteria and promotes soil biological diversity. Michael had to clear some ground on his farm. In doing so, he piled the brush of hardwood trees and deliberately scorched the branch ends and larger stumps. He buried the material in soil, leveled it out, mulched it heavily with ramial wood chips, and planted orchard trees. He believes it to be the healthiest section on his farm.

Hugelkultur can be a great option in areas that have infrastructure below ground: pipes and cables, places where it would be questionable to plant perennial trees with deep root systems. In a place where you've got a lot of downed trees, they can be put to work feeding the soil. Storm-ravaged areas would do well to work with ideas like this. Since Hurricane Irene and the resulting flood that we had here in Vermont a few years ago, we still have lots of downed trees that could be used for this kind of farming.

Here at our home farm, we haven't implemented a full-scale hugelkultur planting as of yet. Though we have used parts of the idea to build up our raised vegetable and flower beds, and have thought about how some of the elements might be put to use in our vineyard, especially the stones to capture heat. And while bits and pieces work well here and there, we are preparing for our own mound experiment.

Below our lower walled garden, which is below the house, there is a gentle slope. It is also an area where there is a lot of infrastructure for the property: leachfield, drainage pipes, plumbing. Right now it's unusable, relatively wild land that gets mowed once a summer. For a long time, we had imagined a series of stone terraces stepping down toward our hop hornbeam wood, but with all the other projects we have going on, stone wall building seems a long way off.

Instead, we've decided to start our own hugelkultur project not unlike what Michael Phillips did in his orchard. This too would be for orchard trees. We'll use felled birch, poplar, and maple that we burn slightly with a leaf fire, then bury them in dirt from another earthmoving project in another area of the farm, mixed with some compost for the topsoil. We'll plant more orchard trees: two or three steps of Cox's Orange Pippin, Frequin Rouge, Old Pearmain, and Normannischen pear.

We've begun to collect the wood that's fallen naturally along our stone walls or that we need to thin out, and hopefully, in a year or two, we'll be able to see the effects of at least one stepped mound with thriving fruit trees.

OUR CIDER TREES

In our orchard, we have focused on Liberty, Cox's Orange Pippin, Old Pearmain, Frequin Rouge, and Orleans Reinette apples, and Normannischen and Barland pears. I love the histories just as much as the names. Cox's Orange Pippin is an apple cultivar first grown in 1825 at Colnbrook in Buckinghamshire, England, by a retired horticulturist named Richard Cox. Though the exact parentage of the variety is unknown, Ribston Pippin is a likely contender. Ribston Pippin also goes by the aliases Essex Pippin, Beautiful Pippin, Formosa, Glory of York, Ribstone, Rockhill's Russet, and Travers.

The Ribston Pippin was a tree grown in 1708 from one of three apple seeds, or pips, sent from Normandy to Sir Henry Goodricke at Ribston Hall in Yorkshire. The original trunk survived for 127 years when the main trunk seemed to die. That season, it put up another shoot, which survived until 1928.

Cox's Orange Pippin is medium-sized, with mottled colors of brilliant red and carmine on a honeyed yellow background. It's very aromatic, sweet, crisp, and juicy. It is a beautiful dessert apple, and an excellent blending apple for cider.

Frequin Rouge is a cultivar originally from Brittany and Normandy exhibiting bittersharp and bittersweet elements good for cidermaking and calvados (apple brandy). It is one of the preferred varieties to go in a calvados blend, offering a clean, astringent note to the sweeter, caramelly flavors of the brandy. As a tree, it is known for being precocious and very productive and is a favorite as a cider apple despite not always being the most resistant to fire blight.

Old Pearmain goes back a very long time. *Pearmain* is an ancient word that was used to indicate that an apple had some likeness to a pear, either in shape or flavor. The direct translation is "the great pear apple." The oldest Pearmain recorded is what was known at the time as Winter Permain or Old Pearmain, which dates back to around AD 1200, at

which time two hundred Pearmains and four hogsheads of Pearmain cider were given to the Norfolk county offices for the Feast of St. Michael every year.

The current cultivar of Winter Pearmain is slightly different. In the nineteenth century, it became confused with the Winter Queening and was sold under the name Winter Pearmain. We can't be sure now that what is labeled as Winter Pearmain is actually the same as the original tree from so many centuries ago, but there is a very strong genetic resemblance between what we see listed as Old Pearmain and Winter Pearmain.

The Old Pearmain appears in Normandy also in the year of 1211 simply under the name Pearmaine. Nursery catalogs suggest the tree came to be known as "old" Pearmain in modern collections starting with that of Alexander Forbes, gardener at Levens Hall Nursery, Kendal, in 1820, who listed it as "Old Pearmain, a dessert apple ripe in November, keeping until January."

Today, there are twenty-four or more apple cultivars that go under the name Pearmaine, with descriptors, or identifiers, like Adam's, Baxter, Claygate, Hormead, Lamb Abbey, Mabbot, Merton, Worcester, Blue, White or Winter, and Golden. I like knowing that there is record of the Winter Pearmain being grown in Indiana in 1800. The Old Pearmain has flesh that is very white, like new snow, crisp and bright.

Orleans Reinette is considered a most handsome and somewhat elusive apple. It has been grown in Normandy for many centuries, with 1776, our American year of independence, being the first time it was recorded in print. In the 1929 book *The Anatomy of Dessert*, the famous English orchardist, nurseryman, and passionate epicure Edward Bunyard writes how the Orleans Reinette

seems to come from the Low Countries, where we first meet with it in 1776. Its brown/red flush and glowing gold do very easily suggest that if Rembrandt had painted a fruit piece he would have chosen this apple. In the rich golden flesh there is a hint of the Ribston flavour, much of the Blenheim nuttiness, and an admirable balance of acidity and sweetness which combine, in my opinion, to make the best apple grown in Western Europe.

It has a flattened shape with a reddish-brown skin and a rosy cheek, firm and creamy-white flesh. It tastes a little exotic, bringing to mind winter oranges and walnuts.

For pears, we have two varieties we grow for cider: Normannischen Ciderbirne and Barland. Originating in Normandy in 1913, Normannischen pears are grown in Normandy and Austria—the German name proclaiming the French roots. The fruits are small, green-gold, and russeted. Barland is also known as Bosbury, and it's an older cultivar of pear originating in England around 1600. The original tree grew in a field called Bare Lands in Bosbury, Herefordshire, in England, and in 1830 it was thought to already be two hundred years old. The blossoms are said to be some of the most fragrant of all pear trees, with a colorful bloom on its skin more like a damask rose or red apple than a pear. They are good for perry because they are high in acids and high in tannins. They make a large, vigorous, and long-lived tree.

Sprays, Teas, Concoctions

Just like in the vineyard, we rely on mineral and plant-based materials for spraying in the orchard if we spray at all. And also like in the vineyard, we only spray when we need to or when we feel the trees will benefit in some way. A lot of the time in the orchard, we let the trees alone and let them grow somewhat untended, keeping our pruning limited and in some years nonexistent. Their hardiness and acclimation to our climate and our place make our trees good growers, easily regulating themselves in relation to the nature of the season.

IN BIODYNAMICS, it is always recommended to spray the 500 preparation, or the barrel compost like what we made with our friend Joseph, in the early spring to jump-start the orchard. There can also be an autumn spraying of the 500 from October until the ground freezes. It settles into the soil over the winter months and gives the orchard an additional medium in which to balance itself.

One of the most important things you can do in the orchard is to use tree paste, which is the best compost for fruit trees as it is applied directly to the "skin" of the plant.

We use tree paste in spring after any winter-into-spring pruning, and if we are extremely organized in our fall chores and it is not too cold after all the leaves have fallen off the trees, we'll spray a thin paste on the trunk and branches before winter really arrives. Tree paste can be applied in a few different ways and in different consistencies for different results. A thick paste for trunks and large wounds can be applied with a paintbrush, and it is easy for the small orchard grower to paint each tree trunk for general health, or for a specific ailment. In larger orchards, time may be best spent choosing only trees that truly are in need, or spraying only one section of the orchard a year. In biodynamics, practitioners recommend that young, year-old trees should always get a good dose of paste, whether thick or more liquidy. A thinner, sprayable liquid for trunks, wounds, and crowns can act as a good preventive.

There are several different recipes for tree paste and different reasons for a thick or a light application. But all recipes share two base ingredients: cow manure and clay. Manure should be clean of straw or shavings and should come preferably from lactating, organically raised animals. The reason for this is philosophical for me, but it is also practical. In applying paste or compost, you would not want to work with manure that is contaminated with hormone growth drugs or antibiotics that will go into your soil. If the only source of manure can come from a dairy farm that does use such drugs for their animals, take manure before the late winter application of the hormones and antibiotics. The manure will be much cleaner at this time, and much more beneficial. Not only do we need to think about our plants thriving, but we should think about the materials we use to encourage their growth—they too should be full of life and energy.

Kaolin is more durable than bentonite (which is often recommended), which can wash off more easily in rain. You can always mix different kinds of clays together to harness their individual properties.

To make a good, fluid paste of cake batter consistency, you can add any of the following liquids: whey or whole milk, horsetail tea, nettle tea, or any other plant-based decoction or tea, horn manure, or barrel preparation. If necessary, different minerals can be added, such as ash, maerl, basalt, potassium salts, diatomaceous earth, potassium permanganate, pigeon droppings, or propolis. Each of these natural substances has different properties for different ailments.

Almost every biodynamic teacher has a special recipe for tree paste. You can find the Australian method, which focuses on filtered cow manure and pigeon droppings, or Ehrenfried Pfeiffer's method, which calls for diatomaceous earth and horsetail. Volkmar Lust's method calls for potassium sulfate and wettable sulfur specifically as a preventive treatment for scab. Maria Thun advocates the simplest of pastes: equal parts cow dung, clay, and whey. There are preventive pastes, boosting and composting pastes, pastes for wounds and for pruning scars.

We use a simple recipe with equal parts whole milk or whey, clay, and manure, with an addition of barrel compost. We make this sufficiently liquidy to spray in both the vineyard and orchard come spring.

And just like in the vineyard, any spray program adopted in the orchard should focus on making the orchard healthy rather than killing unwanted pests. For many growers this can represent a real shift in thinking. Copper and sulfur, or a combination of the two in a Bordeaux mixture, can be used in small quantities just like in the vineyard to combat things like the mildews, but we've found that in our climate and with these fruit cultivars, we've been able to rely solely on plant medicines and compost tree pastes. Other plant boosters, such as liquid fish and seaweed extract along with horsetail and nettle, make for good healthy brews for the trees during the spring to give them a good start to the season if you feel they need a little lift.

Biodynamics calls for a consistent spray program to keep the trees healthy and fed. Permaculture techniques tend to call for less intervention, relying on soil preparation and supportive companion planting to do the work of creating a healthy environment. We use elements of both philosophies and have found them both effective. Biodynamics has an excellent selection of tools, especially when you have problems in the orchard or if you are converting land from one method of cultivation to another.

Biodynamics on its own can be very labor-intensive and take a lot of time if you are trying to follow every single element to the letter. When our vineyard and orchard were smaller, we did follow a strict biodynamic program, and definitely saw the benefits and the results. But when you have as diverse a work life as your farm's plant diversity, a full day's farm work may not be possible every day. Permaculture methodology has intrigued us not only in its foundation in soil ecology but also because it approaches growing from the soil up and asks the grower to focus on the health of their soil and the planting design in order to maximize the benefits of plants that live and work well together. Once you have the design in place, the time involved in maintenance during the growing season becomes less intense. This gives you more time to simply enjoy being on your farm and in your garden.

In the orchard, to combat general fungal disease, our approach is similar to what we do in the vineyard and elsewhere on the farm. A horsetail decoction or tea, tree pastes, and the use of horn silica (known as the 501 recipe), or other silica-based materials like diatomaceous earth or kaolin clay are excellent basic measures to take against fungal diseases like downy mildew and powdery mildew that can cause problems during the course of the season. So far, we have not yet found the need to use these elemental fungicides in the orchard.

Another clever little remedy found in biodynamic methods for the orchard is supplementing the liquid balms with an ash made by incinerating the diseased wood during a certain moon. You grind the ash with mortar and pestle and dry-spray them on the diseased area of land. You can also prepare a fermented liquid with the diseased parts of the tree and then apply that once it is ready. There is a whole school of thought and practice of study, called isopathy, related to homeopathy. Isopathy treats negative with negative, and similar to homeopathy, in dilution. In biodynamics, pathogons are treated with pathogens via the burning, diluting, and spraying of plant diseases and insects. More information about how to prepare these kinds of preparations can be found in books like *A Biodynamic Manual: Practical Instructions for Farmers and Gardeners*, by Pierre Masson. The spine of the copy of his book we have is now well worn, with many pages turned down to remind us of remedies or protocols to follow. It is very direct and easy to understand and puts common sense at the forefront of biodynamic principles.

Come harvest time we are fortunate to have beautiful and very tasty apples from our trees. They are not necessarily perfect apples, but we don't need to sell them at market as dessert fruit, so our expectations are not set for blemish-free. I think this is an important discussion in growing both tree fruits and grapes—to keep in mind as a grower the fruit's purpose. Our goal as wine- and cidergrowers is to have healthy, delicious fruit that speaks of the terroir and tells the narrative of the season. We aim for the healthiest and

cleanest fruit as possible in a given year. We do see things like scab and plum curculio, typical apple maladies, but rather than panicking during the season, we will note the depth of the effect and their existence. We may follow the protocol for an in-season remedy, but mostly we are looking to next year and what we can do to improve the preventive conditions for the trees. We'll look to how we scheduled our work—when we cultivated, planted cover crops, sprayed early spring concoctions. Timing, especially in a difficult season, can be everything, because insects and pathogens happen according to their own cycles, and if you want to break their cycle you have to know when and where to do so. Insects and pathogens are affected by the climate, by the weather patterns. For example, this past year, with inordinate amounts of rain, it was a constant race to keep up with the cycles of black rot and mildew on the property. There are very tight windows for spraying things like copper and sulfur and horsetail and nettle when you are head-to-head with an imbalance on your farm due to weather conditions. If you miss a cycle, it can cause disorder and sometimes the loss of a crop.

Every farmer has to decide what is most important: saving a crop, or saving the plant. Does the farmer look to the farm's immediate economic and health needs or to the sustainable needs of three years down the road? Every farmer must decide if there is a way to blend the responsibilities of the two. Sustainability doesn't just refer to the kinds of agricultural practices a farmer employs on her farm; the farm must be economically viable whether just for home stores and providing food for the family or for more commercial ventures.

No farmer is perfect. Too many chores call on any given day, or other aspects of life must be attended to, and, for certain, the schedule for living will sometimes get in the way of the schedule for farming. The farmer must acknowledge this, do the best he can in the season, and make plans for next year. And, within this kind of situation, the farmer may have to make some difficult decisions. Certainly, no one ever told me that farming would be easy.

Two years ago, the weather was so cold and rainy during the blossoming of the fruit trees, from plums to cherries to apples and pears, that we had no apple crop. Not only did all the pollen get wet, but the temperatures were too cold for the bees to venture out and do their work as crafty pollinators. With our plum trees, we tried brushing the blossoms with feathers, an old farmer's trick to try for pollination, but it was still too cold for the pollination to take. By the time we got to the apple blossoms, which come a little later, we got a frost that withered and browned the tender white flowers and new unfurling leaves. We harvested two apples that season. I've never been so thankful for our farm's diversity.

Although the fruit trees suffered, the weather cleared and dried up by the time the grapes and other fruits and vegetables blossomed. We had a solid wine harvest, we were rich in tomatoes and Delicata squash, had heavy stores of French green beans, and we had a bumper crop of the most stunning daikon radishes we've ever grown. All these crops fed us personally, and they went into the kitchen at our restaurant. They not only gave us life through their bounty but also provided us a living.

Almost exactly the opposite happened this past growing season. We had very warm weather during the blossoming of the fruit trees, then long periods of cold and rain for the first half of the summer. We had just enough heat and light in mid- and late June to foster the blossoming of the grape vines. While we had excellent fruit-set, one section of our vineyard suffered significantly due to black rot, a result of all that early wet weather. This was one of those times when life's schedule got in the way of the cycle of my least favorite fungus. As a consequence, we ended up having to drop a large percentage of grapes in that block in order to save the plants for next year. Yet the rest of our home vineyard thrived, and we still had our largest harvest to date.

The rain this past year caused all kinds of confusion in the other gardens. Our greens were abundant, lush, almost obscene, while it took forever for the tomatoes to set fruit and to eventually ripen. The beans were all late—they had to be started three different times. We had another dry and drought-driven latter half of summer and a spectacularly warm and sunny fall. This boded well for the vineyard and those tomatoes. In difficult circumstances, all is not always necessarily lost.

To Prune, or Not to Prune

Some years we prune, some years we don't. Sometimes it's a matter of orchard philosophy, sometimes it's a matter of time. Classic orchardists recommend pruning, telling you that it is the best way to control the vigor of the tree, to encourage blossoms and therefore fruit, to keep the tree open to light and air by cleaning out unruly limbs that cross and block, to take back those watersprouts, fast-growing whiplike branches that rise toward the sun. There is a long tradition of the shaped and cultivated tree, torquing and gnarling thickly as it gets older and builds a callous of bark from each pruning season, looking much like the broad, twisting arms of an older grape vine that has also seen season after season of cutting back.

A PERMACULTURIST WOULD tell you to avoid pruning as much as possible, especially in a snowy climate like ours, where heavy, wet snowfall can burden the younger, weaker fruit tree branches that grow in response to pruning, to the point of splintering and breaking them in a dangerous way for the tree. A permaculturist will point to the tree in nature, citing the fact that they find

their own balance without the help of human hands and that the tree will regulate its own vigor and its own ability to produce a healthy harvest in a season. The tree knows better than we do.

Both methods share good reasoning. And it is certainly more difficult to let a tree go unpruned once it has experienced pruning. When those first cuts are made, the tree responds. The tree will try to grow multiple branches around the pruning wound site—these are the watersprouts. The watersprouts have to be removed, otherwise they will take energy away from the tree, frustrate its intent to produce fruit, crowd and confound good air circulation within the canopy, and make the tree more susceptible to fire blight infection.

But if a young tree is left to its own devices, it seems to monitor its own growth patterns. When I look at the wild trees on our land that don't get pruned, I see how this works. These trees are not crowded with branches; they have an elegant, natural shape. They bear a good amount of fruit but are balanced, not overabundant. They may search their way around other trees and bushes in a hedge in order to see more sun, which will shift their shape, but there is nary a watersprout in sight. There is something real to think about here.

When we began with our apple trees, we began with pruning. And just as in the vineyard, there is a very real satisfaction that can be felt when grooming a tree and giving it breathing space. Early on in this narrative of our farm, I said I never expected to be a farmer. I'm not entirely sure what I expected— maybe days at the head of a classroom as I once thought I would be a teacher, or in a studio painting—but I didn't know that working the land would offer some of the most grueling, heartbreaking, and satisfying work I've ever done. I didn't know that the work would actually feel like a second skin, almost like intuition. And that's the strange thing: We are still so young in this work as farmers, and there is still so much to learn and understand, that when this farming business feels absolutely right, it takes you a bit by surprise.

Learning to prune felt like that to me. I remember well the winter it happened. The sun was beating down, and this was one of my days to prune in the orchard. Caleb had pruned all the week before. I strapped on my snowshoes and had one set of pruners in my back pocket and one in my hand. The light bounced brightly off the white landscape, so much so that I wore sunglasses. Our trees were big enough to be generous bearers of fruit already and usually much taller than me. However, with 2 feet of snowpack, I stood that much closer to the tops. But I'd still have to use loppers to reach the highest points.

Before that winter, I'd never pruned our trees, other than occasionally stealing flowering branches for a bud vase or a floral arrangement in the osteria. This is in and of itself bad behavior. To prune during flowering is like tying your dog on a very short chain in a dirty yard with no shade and no water, or forgetting your child in the grocery store and driving on home without a care in the world. I realized as I looked at those trees I would have to stop that behavior. At least on our own trees. I could pilfer elsewhere. But that's always been my modus operandi: Rosebushes, hydrangeas, and peonies in other yards are never safe from my covetous eye, and my snappy sheers.

Somehow the real, legitimate pruning of our trees had always fallen under Caleb's purview before that snowy season, and because his days were already quite full, it hadn't always been easy for him to finish the work. That year, we'd decided to get really serious about the apple trees, so we were sharing the work. At the time, I was already feeling nervous about pruning our grape vines come April, even after the patient tutelage of our friend Emmanuel in Burgundy in the autumn. Pruning grape vines can make or break your plant, it can be the deciding factor between a good season and a bad one. But apple trees are not the same creatures.

I'd taken some time to look at our handy *Little Pruning Book: An Intimate Guide to the Surer Growing of Better Fruits and Flowers*, by F. F. Rockwell, published in 1919. Mr. Rockwell has many good things to say about the process of pruning, but he has four points he says to be sure to always keep in mind, and which I always carry with me at the ready when pruning, just like the extra secateurs in my back pocket:

FIRST: *Always leave a clean smooth cut.* Careless cutting or dull shears, leaving a ragged edge, means slow healing and increased danger—to say nothing about its being the earmark of a slovenly gardener.

SECOND: *Cut just the right distance above the bud.* If you cut close to it, it is likely to be injured. If you cut too far above it, a dead stub will be left. On small branches and twigs, cut from ¼ to less than ½ inch above the bud. If pruning is done when plants are in active growth, however, the cut should be made close to the bud, as it will heal almost immediately. (Note to self when pruning other people's trees under the cover of darkness . . .)

THIRD: *Prune above the outward-facing bud.* This will tend to keep the new growth branching outward, giving the plant an open center with plenty of space and light. While in some specific cases there may be reasons for selecting an inside bud, this holds as a general rule.

FOURTH: *Cut close up to and parallel with the main branch, trunk, or stem.* In removing a branch from a tree or side shoots from shrubs or plants, the leaving of a stub, even it if is a short one, delays the healing or makes it possible for disease and germs to enter, thus providing for future trouble.

S o, with a fair amount of trepidation, I started, that bright sun beating down relentlessly on a cold winter day.

I TOOK THE PRUNING branch by branch. I stepped away occasionally to look at the tree as a whole, asking the question: Are the branches balanced? The work went both slowly and quickly. There is a meditative quality to the process, and time seems to be neither moving nor standing still. The sun was warm and brilliant and felt like a balm to cold bones. The air smelled fresh and shivery, and felt like it must be full of the best oxygen. After a while, I realized I was already on the third tree and any residual fear was gone.

The snow-covered ground was littered with fallen branches to be collected in bunches. Some came inside to be forced for blossoms in vases (old habits die hard); others were evaluated for suitability as possible cuttings; still others were left to dry as wood for cooking. (Doesn't duck roasted over apple wood sound pretty delicious?)

The trees looked airy and shaped liked lacy goblets, arms reaching out and up. When Caleb returned home, he helped me reach the tops I couldn't quite get to. The sun started to shift. It was already three in the afternoon, and we had yet to eat lunch. We decided to stop for the day and catch a bit of sun on the porch with a glass of wine, some salame, little pickles, and bread. We closed our eyes to the heat on our faces and the thought of bees humming in blossoms in just a few months' time, and then a few months after that, the fruit ripe on the trees would drown out completely our winter scene.

But the theory that a plant can take care of itself is also very appealing. We've thought about this untended sensibility in both the vineyard and the orchard. It can be rough going the first couple of years, but vines and trees—in fact, all perennial, woody plants, according to others who work in this way—should find balance on their own within three years. I have seen it in other vineyards; I have seen it in the wild fruit trees surrounding us.

Our friends in Austria who leave one of their vineyards untended have proven this theory with some of the beautiful and idiosyncratic wine they produce from it. I think about this for our vineyard; I think about it for our orchard: I love the idea of even going back in time and growing vines up trees like the Romans did, maybe even apple trees, for support, but without pruning. We have experimented a bit in the orchard with this, and as well in the vineyard. While the orchard has been a planned examination, the vineyard trial was born out of circumstances. An unexpected loss of hands for pruning in two vineyards where we previously bought fruit (but now lease and farm) left several rows of vines to their own devices. In some of the rows we pitched in and tried to bring the plants back under control, as many of them had not been pruned the year before, either. Our northern hybrids are extremely vigorous and will voraciously eat real estate in their rows if allowed. Single vines had cordons and attendant tendrils winding and snaking into plants three vines down. It was bit like a *Gray Gardens* vineyard, or what might look like *Great Expectations'* Miss Havisham's version of a vineyard. There was a very appealing wildness to these rows, and the amount of fruit thrown by each plant was awe inspiring. But ultimately, I think our climate, and our microclimate in these valley vineyards, may be too problematic to grow vines this wild for the health of the plants and the fruit. When working organically, it is important to be able to keep the plants open to light and air and not trailing on the ground, to keep the mildews and black rot at bay given our humid and often wet early seasons.

In these wild rows, the cluster quality also suffered, even though many individual berries were spectacular throughout each vine. The cluster shapes were long with extended space between berries, and while ultimately that was good for air circulation, it made it very difficult to harvest effectively.

Our hillside home vineyard may be a different story. Our altitude may curb these plants' enthusiasm a bit, as well as the close spacing between vines we have chosen. We certainly have wild grapes that crawl and climb all over trees in the hedge. A new experiment with our wild apple trees has started to take shape. Vines will be planted at the base, allowed to grow unpruned. We will choose the St. Croix vine as the first candidate, as it is extremely hardy and disease-resistant. There are so many intriguing questions to be answered, or even just posed. What happens when grapes and apples grow together? A cellar geek would probably warn me about the different yeast cultures on the two different fruits, but we have also done some experimenting with cider fermented on the lees of our red wine, and this might be an interesting companion planting for that cider.

In many ways, the apples and the grapes are already entwined in my approach. On the home farm, they reside side by side in their permanent and perennial culture, and while there are some differences in how we address their cultivation, many of the ways are the same, and we intentionally tie those ideas together for ease of care. The wines and the ciders also ferment side by side in the cantina. My philosophies in the cellar are the same for both wine and cider. Again, a cellar geek might tell me that I could encounter complications between the different skin yeasts and fermenting by-products; in fact a cellar geek *has* told me such a thing. But our efforts are so simple, and rooted in cleanliness more than anything else, that we have yet to see any issues arise. The fermentations generally take place at different times, as does the bottling. Not only do they live happily side by side in the field, but they do so also in the bottle.

This duality between the orchard and the vineyard points so clearly to how we grow things on the farm in every way. The duality is actually a plurality. While I've broken the farm down into separate landscapes, or chapters, it is difficult to write or talk about one without the other. They are all so interconnected, and one aspect of the farm feeds and informs the other. Many of the vegetables we grow are cultivated in the vineyard, or beneath the fruit trees, or alongside cuttings in the greenhouse. The roses and flowers find their way into the vineyard, and are mingled with vegetables and herbs. The bees, our only nod to "livestock" at the moment, are in every element of the farm. They pollinate and take nectar from every garden and every plant.

They in turn produce not only honey for our table, but also future propolis for our plants.

One piece of the farm doesn't exist without the other. The finely wrought web and the gossamer chains connecting each of these moving parts amplify and shape. Like a narrative or an assemblage, the whole is greater than the pieces. It is what makes this small plot of land cultivated with nature so vibrant and full of perfectly imperfect life.

Walled Garden

Much like I longed for a little house in an orchard, I always desired a walled garden. Walled gardens evoke great houses with extensive kitchen gardens; forgotten and secret gardens; wildly romantic naturalistic gardens; enclosed private, quiet, and contemplative gardens. Somewhere along the way, I became entranced by the idea of the sunken walled garden . . . an adaptation of an old stone cellar hole, the house long gone.

I HAD SEEN such a cellar hole near my parents' home. The old stone foundation dug deep into the earth as well as reaching slightly above the grade of the land. One wall had collapsed, allowing the passerby to see something of the inside. It anchored the center of a wild meadow that no longer saw cows or sheep to graze it. The stones were dark and gray, round and oblong, irregular. Typical of the native New England stones that come up from the ground every time you plow or hoe, not unlike knobby potatoes in winter. The stones were covered in lichen and moss, the colors of verdigris and tree bark. The field was overgrown; the cellar hole, too, overgrown, a forgotten remnant of an old farm. Queen Anne's lace, black-eyed Susan, dock, fragrant bedstraw, and wild sorrel grew rampant between the walls. A matted-down path wended its way in and around the stones, probably worn by a single deer searching for dinner. A twisted and knotty apple tree hung over the edge of the wall, in the spring the blossoms profuse and pink, and in the fall the branches heavy and drooping with celadon-colored fruit.

This sunken garden captivated my imagination like a good story or a mesmerizing painting. It was a marriage between human structure and landscape, the wall and its bevy of wild flowering plants sitting snugly in the meadow, human architecture accepted by nature.

I wanted to know the story of this garden. Where had the house or barn gone? Had it been destroyed by fire? Had it fallen into disrepair and been taken down? Had it been abandoned once before, like it so clearly was abandoned now? Where was the family that had lived there? And if I couldn't find out the exact history of the place, my imagination wanted to create it.

I haven't been by that completely wild and natural garden for many years; my parents no longer live nearby. But the image is crystalline as if I had just seen it, or even as if I am standing in front of it. Every detail in each of my senses is etched, even the summer fragrances and the sounds of the crickets attendant. And while I know that I am not there, and nowhere near there, that garden, even as I think of it at this moment, is completely present.

When we finally bought our property, it never occurred to me that one day we might have a walled garden, let alone a sunken walled garden anything like the one that had so firmly planted itself into my memory. Just as there were so many other things missing from the land, or so we thought, like no apple orchard or climbing roses, there was no abandoned stone cellar hole full of possibility and promise.

Confronted with so many other projects calling in our new home, fancies of stone gardens disappeared. It wasn't until the original garage was gone and the new barn built that the empty space between the road and the house, an awkward slope that delivered unwanted rainwater runoff straight into our own cellar, revealed the apparition of the sunken walled garden. And once the apparition had appeared, there was no exorcising it. Nor did we want to.

It was midwinter when that spirit of a garden came to us. We knew we could do nothing about realizing it until spring came, but the garden haunted that foggy, snowy winter landscape. It beckoned coyly with promises of exuberant and wild vegetation. Roses certainly, a rambling grape vine, herbs, and wild lettuces. An apple tree that blushed prettily in spring and fed in autumn. It was just like when we were bewitched by the house and land before we moved here.

WALLED GARDEN IN AUTUMN

Finally, spring did arrive. Our neighbors down the road came with their tractor and helped dig out the earth from the slope, depositing it up above and flattening it up on the driveway at the same level as our road. With the huge gouge in the ground, we were able to procure help from two other neighbors who specialized in building stone walls. We wanted all the stone to come from our land, from the fallen stone walls that lined our property. Stone walls once built by a farmer to protect his sheep; stones that percolated up from our clayey, silty soil; round, soft, gray stones; quartzite and thick, flat pieces of rust- and sand-colored shale: the same stones that would have built the foundation of a barn or a house two hundred years ago; the same stones that did build the foundation of the barn belonging to our neighbor, the barn that used to belong to our property. Or perhaps it is more accurate to say our land once belonged with the barn.

It took two years to build the sunken wall, which is terraced in three places as it descends to our house. By the time the walls were finished, the

bed of the garden was uneven and overgrown with thick mats of crab- and spider grass, and clumps of goldenrod. The floor of the garden was definitely wild, but not the pastoral magic that I had envisaged. Already it was the end of the summer, a hard time to do much significant work on the space when so many other things were happening. One of which was the planning of a friend's wedding that was to take place in a couple of weeks at the farm—I say farm, though I am not sure it was a farm yet then. We only had the one large garden and one half of the orchard. We had a beautiful meadow that we had been reclaiming from hungry poplar. But I do know the intention toward farm was there.

The walled garden was a perfect place for a dance floor with a tent over it. We could play music from the guest bedroom window that fronted on the space. I spent a backbreaking week forking up the dirt, cleaning out the grasses and goldenrod, trying to level out the ground. We laid straw down over the edges of the area, and the day before the wedding Caleb built a plywood dance floor on risers in the middle. That day before the wedding, the rain and surly winds also began. The tents went up as the weather worsened, the fabric and poles anchored tightly to the ground. As the weather continued to turn toward dark and menacing skies and spitting rain, the dance floor would become the reception area before the dinner under another, larger tent. We knew it might need to also be the space for the wedding ceremony itself if the weather prevented it from happening at the edge of the field under the sheltering branches of an ancient copper beech, which is what the bride wanted.

The stone wall embraced the white tent. White paper lanterns blew in the rain and wind and, as the sky blackened even more from the heavy cloud cover, the tent glowed in the dying light. So the origins of our walled garden began with a wedding and a dance. Per the bride's request, the ceremony ended up in the field under the beech tree, guests protected by so many umbrellas, but the first toast was made in the walled garden, and the first dance was trod on the makeshift floor. While the couple was congratulated, we all shared in food and drink, celebrating love in their lives as well as our own.

After the dinner, the rain stopped, fog wreathing around us in the black night lighted by candles and sparkling lights. The music came out of the guest room windows facing the walled garden, and the dancing began. A

WALLED GARDEN WITH BARE BEDS

long evening of music and energy, laughter and release. An auspicious christening to the garden.

After the wedding, we did manage to do some serious work on the garden. Tilling and leveling, rolling out a flat space between the gracious curve of the walls. We bought four crabapple trees and evenly placed them in a square at the center of the space. Their dark red berries were bright against the fall gray skies, and the trees added an elegant dimension to the unfinished space. We buried them in their pots partially in the ground for the winter. Then we spent the next few months envisioning and planning.

While I had this hope of a wild garden, Caleb had the vision of a traditional French potager, or kitchen garden. I was not opposed to this; in fact I embraced the idea too, having spent so much time admiring and coveting potagers that we had seen while traveling. But I still wanted the effect of the

WALLED GARDEN IN MIDSUMMER

walled garden married to nature, a seamless expansion of what nature and the human hand had to offer together.

This was the first real effort Caleb and I had worked on together in terms of a landscape design, an understanding that what we wanted to do with our land could be created and designed in a fashion that brought our life and our natural world together in a way that worked—the meaning of this twofold. Of course, we wanted the design and implementation of it to work aesthetically and physically, but we also wanted it to work as in productively "working the land." We agreed on traditional raised box beds for the plants, and on small peastone paths for the spaces around the raised beds. We agreed on the stone wall providing a backdrop for a more wild expression and experimentation. We agreed on the crabapples anchoring the four interior beds.

When spring came, Caleb could hardly wait to get into the garden. Raised boxes went up, eight of them, made from the pile of remaining cedar lumber from the old garage. The beds were filled with composted soil, the result of compost that we had been making from the collected refuse at the osteria. The dark and fluffy soil was ready to accept the first plants. This is where our smooth agreements parted, at least for a moment. In my mind, I had imagined the beds lined with rectangles of boxwood holding wild roses and the arching stems of daisies, maybe round purple and green cabbages or spiky artichokes. I thought we could have a mix of flowers, fruits, and vegetables. Caleb had a different set of plants in mind, all of which tended toward the edible. It took me a few days to change the direction of my walled garden fantasy. The reality was that there just wasn't enough room for it all within our narrow beds, built with the ease of cultivation in mind. By the end of the week, I had come around. The thought of a true potager with tomatoes and lettuces and a trellis for French green beans growing was just as seductive, if not more so. We came to another agreement: The beds would be reserved for vegetable growing, but the perimeters and the upper terrace would form a blank canvas for wild flowering plants and bushes.

As we began to look at the space as something we would shape, wildness gave place to structure. The raised beds, along with the fruit trees and the shape of our stone walls, already provided some architecture, but the more we learned and studied plants and landscape, the more we saw the beauty of rhythms in structure. Structure is a foil for wildness. I became interested in further explorations of ruins in nature—that initial stone cellar hole the ruin that started it all. The idea of structure abandoned to nature was extremely appealing, especially at the moment where nature and architecture are completely balanced, before nature completely takes over. The question was how to achieve that moment of balance not only with the design, but with the plants as well.

This is the same impulse that drives my interest in the untended vineyard, the untended orchard. The balance between structure and nature, human hands and nature. And can that balance be perennial? Every time we grow something new on the farm and we create a new space for planting, we are confronted with this question of structure and nature. Aesthetically, structure provides the backdrop for wildness, the bones: it supports, contrasts, pushes you, and pulls you. And it does the same in terms of function. I sense this will be a lifelong question and experiment. Sometimes I feel we have added too much structure, and sometimes I feel we have too much wild. I think I will always be grappling with this tension between the two as long as I have a landscape in which to plant myself.

The Garden Extant

Choosing a garden space for vegetables isn't as crucial as it is for a vineyard or orchard. You can garden almost anywhere as long as you work and amend the soil if necessary and you either have access to water, or grow plants that require little water. Vegetable gardens are enormously kind: Our first two grew on their own with very little help from us. Of course, you ideally don't want to plant your garden in a swamp or a desert, but if that's all you have, you can build raised beds, compost, and choose plants wisely—edible plants that will grow in those circumstances. In dry climates, many herbs do well, such as rosemary and thyme, and planting other water-efficient plants at the right time of year can really make a thriving garden; or, in wet or swampier areas, blackberries, melons, asparagus, cattail, lotus root, and other exotic foods can grow swimmingly (the pun really not intended). It may not be the garden of your dreams, or maybe it will be, because what you will plant there is all you dream about. Planting in an area that might not be the classic standard for an edible garden can actually challenge the gardener in interesting ways, urging you to think, and cook, in a new way, one that is appropriate to the environment and your personal landscape.

THERE IS A LOT to be said for the raised bed of the potager as a way to plant a garden. When you build a raised bed, you can grow food in any location— even that desert or marsh. Soil conditions are no longer an issue because you've raised the bed off the ground and brought in compost to build up the present soil. Soil compaction is less of a problem because you are typically not walking in the beds; they are designed so that you can work at their side and reach across to the middle. Weeds are less of a nuisance because the compost has been so well worked. Plus, the crop or crops planted in the raised beds colonize the square footage, outstripping any weeds that might want to grow there. What weeds do grow are easy to pull out because the soil is so loose. The soil tends to drain better. Raised beds create their own micro-climate, warming up a little earlier in the spring, which allows a head start

WALLED GARDEN IN LATE SUMMER

PLANTING ESCAROLE

on spring plantings. Watering can be controlled in a smaller area. When planting directly in the ground, any watering can become diffuse, absorbed toward other areas not near the plants needing a drink. This is not the case with contained raised beds. Because of all these factors, raised beds can provide an immensely productive garden, growing more produce per square foot than in traditional crop rows.

Most gardens need water, and it is important to know where the water for a garden is coming from, and how to manage it in a responsible way. Even if there is access to readily available water through a spigot attached to a house or barn, it's a good idea to think about collecting rainwater. We use a rainwater barrel that catches runoff from our house roof. We also have 5-gallon buckets lined up in a neat row below the shed roof of our barn, which also collect rainwater from the roof. We always use water from the barrel and buckets first. We hope that we never have to turn on the spigot from our well. But it's rare to have a perfectly balanced season in the garden, one with just enough rain to keep all plants happy between rainfall and rain barrels and sun. And not all plants require the same amount of water, further complicating the dream of a perfectly precipitated summer. The nature of the soil—its absorbency and ability to conserve and hold water, and the amount of compost (organic matter) it contains—also plays a role in how to approach water usage and retention in a way that is not wasteful or detrimental to the local environment.

There are countless reasons why raised beds are good, but they also represent an investment in time, energy, and sometimes money. To build a boxed bed, you need wood and nails. Having said that, the box is not necessary. Many organic gardeners simply mound up the dirt to form unsided raised planting beds, but compost-enriched soil is still needed. Creating homegrown compost is fairly easy by composting home kitchen wastes. If wooden bed boxes are to be used, they will eventually need to be replaced, and they can't be moved once they are filled with dirt. The soil must be amended regularly to keep it rich in nutrients and soil life. Raised beds are inherently small, or at least narrow, which can feel limiting for larger crops like corn, squash, and potatoes. But I think these considerations are minor compared with all the advantages.

Soil for the vegetable garden must be full of nutrients. Just like when planning a vineyard or an orchard, a soil test can be a good idea. You want to know the amount of nitrogen, phosphorus, and potash in your native garden soil, as well as its pH, or relative acidity. A soil test can be done, just like for the perennial crops, through the local agriculture extension, or a home test kit can be found through many garden centers. La Motte test kits have always been recommended to us, and they seem to work more accurately than smaller kits.

Testing composted soil for the garden is a little different than understanding the soil in the vineyard or orchard. We rely on what we see growing around the vines and fruit trees to understand the soil's needs. It's more difficult to do that with a raised bed as you are weeding as you go along, you must pay closer attention to what you weed.

Vegetables tend to grow best in a slightly acidic soil with a pH of 6.5, but if there is good organic

matter available, the plants will tolerate a soil that is between 6.0 and 7.5. The plants tend to be more forgiving when provided with all that humus. To lower the pH of soil, amend beds with pine needles, oak leaf mold, decomposed oak or pine shavings, and more compost. To raise the pH, it is often suggested to amend with lime. The kind of limestone seems to be important here. Many organic gardeners and books recommend dolomitic lime, but this is very high in magnesium. If the soil test shows that the soil is very deficient in magnesium, then this could be just the thing. But many soils are not that deficient in magnesium. Calcitic lime is another agricultural lime, which is richer in calcium than in magnesium, and it is often recommended because it appears to be faster acting and doesn't raise the magnesium in the soil to unbalanced levels.

An even better way to adjust the acidity of the soil would be to plant a cover crop for a season or part of a season. Just as in the vineyard and orchard, the decomposing plants change the balance of the soil and can add sweetness, raising the soil pH.

Composting in the garden is the key; in fact, it is key in every growing aspect of the farm. There is composted manure, composted food scraps, biodynamic compost, forest-edge compost. Compost is simply the ultimate act of death and rebirth in nature. In nature, living things die, and their death creates new life. Plants and animals die in fields and on forest floors, and are composted naturally by time, water, sun, air, and microorganisms. In time they become one with the soil, improving it in terms of texture and nutrients.

Compost, Again

Compost comes up again and again in a variety of ways on the farm or in the garden. There are so many different approaches to compost and so much study that has been done on compost that we can see that certain composts help certain plants more than others, and, given that knowledge, why not use particular methods, or particular types for certain situations?

General, all-around compost is good for all plants, but as we know in the vineyard, certain kinds of animal manures are better than others in certain microclimates and for the vines, and in the orchard, composted ramial wood chips are particularly beneficial for the fruit trees. I wouldn't use the ramial wood chips that we've employed in the orchard in the vineyard, just as I wouldn't use all manures in the vineyard. Cow dung mixed with a little horse manure is the best elixir for our vineyard situation.

COMPOST IN THE NATURAL WORLD is created in three different ways. Manure, as we know, is created by plant and animal foods composting inside the body of an animal—earthworms are included here—and then the composting process is continued outside the body of animals in a kind of natural fermentation. I firmly believe that the first marks of civilization are fire and fermentation, and the first signs of life are also fermentation. Composted manure is a type of fermentation that gardeners of all stripes apply to their soils. Understanding this basic function goes a long way toward explaining how and why biodynamic compost inoculants are and were created. The fermentation of certain plants inside certain animal membranes recreates a natural phenomenon. But these preparations are specific, pointed, in order to tease out particular effects and properties from the plant fermentations. General animal manure is a gestalt of the process, whereas the biodynamic

preparations are a "deconstruction" that distills certain attributes to add to your compost and amplify elements.

Earthworms are one of the greatest composters, which is why so many home composters add earthworm colonies. Earthworm castings are five times richer in nitrogen, two times richer in exchangeable calcium, seven times richer in available phosphorus, and eleven times richer than the available potassium in the soil they inhabit. Their presence naturally in your soil and compost is an excellent sign that your soil is working to its utmost. Their addition to poor soil can help balance and enrich. It is no wonder that the full moon that happens near the spring equinox is called the Worm Moon, because in many places, when the snow melts and the earth thaws, the earthworms rise to the top of the soil and slough off of their winter coats, or castings, creating the first compost of the season in the natural world.

Compost is created not only in animal manures but also by animal and plant bodies that decay on top of the soil. One of the composts we use here at the farm is made from farm animal mortalities from other local farms in our area. I've always understood that animal proteins are essential in compost. However, many organic gardeners who are vegetarian or vegan won't use any kind of animal matter in their gardens. Purely plant-based composts can be good, and can be really good for certain plants, but they are only effective to a certain point. Decomposing animal proteins in a mixed compost completes the full circle, providing micronutrients, and once completely decomposed, they provide a big boost of nitrogen. Composted animal matter significantly ups the feeding power of your dirt.

Many gardening manuals shy away from composting animal waste from the kitchen, scraps like leftover meat, bones, and eggs or eggshells. The reasons for this are that such scraps are smelly and unsavory to our noses, plus they are tasty to all kinds of scavengers: raccoons, buzzards, rodents, even bears. And while decomposing meat is stinky, it is also very slow to break down. Vegetable matter decomposes with aerobic bacteria, working fairly quickly and creating a lot of heat. Animal matter decomposes with anaerobic bacteria, which function best without air. But the slow breakdown can attract flies and their fly eggs, which become maggots. Maggots make most people squeamish (myself included), but maggots are some of the cleanest

larvae around, and the work they do is cleansing. Maggot wound therapy has been used since antiquity; the Mayan and Aboriginal tribes used it, and military physicians throughout the ages have employed it in the field. Allied army doctors during World War I and II documented significant benefits to both wounds and bone fractures with maggot therapy. The larvae are used to help clean wounds and initiate healing. The point here is that maggots in animal protein are just cleaning up and preparing for the next phase of decomposition; they are not a sign of dirty compost.

While we get our farm mortality compost already broken down into dirt, for the farmer who has livestock or roadkill that needs a final resting place, these animals can be added to the compost; they will add to the natural bounty of the nutrients. It is a relatively easy and environmentally sound practice, a completely natural way to dispose of animals that have died. In organic and biodynamic farming, one of the goals is to have as few, if any, off-farm inputs as possible. Adding farm mortalities to the compost brings that cycle of life on the farm into a complete circle, and can enhance the fertility of the farm or garden greatly.

There are a few things to do when adding such animal protein to a pile; the same goes for any kitchen waste, such as the bones and refuse that we process from our osteria and home farm kitchens. Some gardeners and farmers recommend a sprinkling of lime to help lessen the smell and speed up decomposition, but straw can also be used, or other green matter, over the animal waste. This will help lessen the smell and the siren's call for scavengers. Turning the compost fairly regularly will also help in speeding up the naturally slower process of animal decomposition.

For a commercial small farm, or for those who produce food commercially from a small garden patch or open a farm stand every season, or for those who make cheese or process canned food or preserves, the state health regulations concerning animal matter in compost should be checked. Every state has different laws and restrictions, and because of the existence of mad cow disease, some states fear the proliferation of harmful bacteria on the farm. There is also currently legislation in the lobbying and review period concerning the application and use of compost on small farms. It's important to check recent USDA regulations and considerations in order to be informed and to make decisions with these things in mind.

The third natural way of composting is through the decomposition of plant and tree roots, root hairs, and their interconnected microbial life forms that decay naturally in the soil itself after harvesting.

Compost performs two functions in the natural world. It improves the structure of the soil and makes it easier to work with. The ground becomes well aerated and retains water; it becomes more resistant to erosion. Compost also provides food for plant growth. Humic acid is one of the primary components of compost, and this makes nutrients existing in the soil, though not always in readily available forms, more readily accessible to plants.

Biodynamic compost with plant-based inoculants is one of the most effective compost preparations to be found. But this is not the only way to compost in what is considered a biodynamic fashion. In studying vegetable gardens and the best composts for vegetables, we have often used and returned to one of the gardening books we found early on in Caleb's mother's bookshelves. A classic book of a certain age by John Jeavons, *How to Grow More Vegetables: A Primer on the Life-Giving Biodynamic/French Intensive Method of Organic Horticulture*, is a trove of very specific information on how to pull together raised beds and create a beautiful and productive garden space. Jeavons was very influenced by the methods of English master gardener Alan Chadwick, who brought these ideas of biodynamic and French intensive gardening to central California in the 1960s.

The biodynamic French intensive method of composting is different than a straight biodynamic preparation and it is what we generally use in our vegetable gardens. There are no added plant inoculants. The recipe is by parts: 1 part dry vegetation (for example, hay, leaves, straw), 1 part green vegetation and kitchen wastes, and 1 part soil, though if you have heavy clayey soil, this method will work better with less soil in the mix. In our flower and herb gardens, we have a heavier clayey soil in some of the beds and have found we can't do enough to lighten the tilth.

FRENCH INTENSIVE

French intensive gardening is a system much like any other long-standing organic method: The goal is for the gardener to work with nature as opposed to against nature and to encourage healthy and fertile plants in smaller spaces with less water than more conventional gardening uses.

The first documented French intensive garden was started in the 1890s on 2 acres outside of Paris as a way for working-class French with small backyards or allotments to be able to grow an abundance of food in a small space for their families. The crops were set in 1½ feet of composted horse manure and were planted close enough together that the leaves of the mature plants would touch one another.

The main focus of the French intensive method is the bed preparation. The beds are created by double digging—layering a compost (traditionally made from horse manure, because that was readily available in 1890s Paris) onto the topsoil and then digging trenches around the perimeter of the beds to loosen the soil beneath the beds. The dugout topsoil gets re-incorporated into the new bed. The suggested size of the beds is about 5 feet in width, so

they are easy to work across without stepping in the beds. The digging of trenches and incorporating that topsoil into the beds creates raised beds with better drainage and well-aerated soil.

By placing the plants so close together, the mature plants create a kind of living mulch across the soil, keeping unwanted weeds at bay. Companion planting is also an active part of this method. When you plant two or more varieties that grow well together, they can increase each other's productivity while aiding pest control, facilitating pollination, and providing a beneficial habitat. Companion planting is nothing new: It has been used for centuries in English cottage gardens and home gardens in Asia, and for thousands of years in Central America. As far back as ten thousand years ago, indigenous Americans began forming a companion planting technique called the Three Sisters, an agriculture based on the relationships among corn, beans, and squash. The cornstalk served as the trellis for the beans to climb, the beans fixed the nitrogen, which benefited the corn and squash, and the squash spread along the ground, blocking out light and preventing the growth of un-

wanted weeds. The squash leaves also acted as a living mulch creating a microclimate that retained moisture in the soil, and the prickly surface of the squash vines deterred pests. Not only does the com- panionship work in the agriculture, but it works on the table as well. The combination of corn and beans pro-vide a perfectly balanced diet in which each plant offers amino acids that the human body needs to survive.

Alan Chadwick's introduction of the French intensive method paired with biodynamics in the United States makes for an interesting hybrid horticulture for the gardener. Chadwick, an actor by training and desire, had also studied biodynamics under Rudolf Steiner and then the French intensive method in France after the two world wars. His approach to sustainable farming was entwined with his hope to help prevent further war by educating young students in their relationship and primal connection with nature and the earth. He inspired scores of young gardeners and farmers to work with the land in conscious, thoughtful, and bountiful ways. He is still inspiring.

The ground beneath the pile should be loosened with a fork 12 to 24 inches down in order to expose the bottom layer to good bacteria and organisms in the new compost as well as making good drainage. Ideally, you want to build your pile from the bottom up, starting with about 1 to 2 inches of the dry vegetation, followed by the same amount of the green vegetation, and finishing with about ¼ to ½ inch of the soil. In a perfect composting situation, you would repeat these layers in this order, but you can also build from these first three layers by adding whatever materials you have daily. Every night or two the osteria is open, we bring home a 5-gallon bucket of kitchen wastes. Every morning this gets added to the compost. We try to always add a little bit of the dry vegetation on top of the food wastes in order to cut down on the flies and intense smell during the warmer months.

The green vegetation is much more effective at getting your compost going than the dry vegetation because the green has a higher nitrogen content, which helps jump-start the fermentation process. Dry vegetation has a higher carbon content, not unlike the slightly burned logs used in hugelkultur or biochar. Carbon is an excellent resource for your soil and plants, but it is more difficult for your compost pile to process carbon-rich material without a lot of nitrogen already present. In a smaller household, like ours, it can take several days to accumulate the proper amount of kitchen waste material for that "green" layer. We keep it in securely lidded buckets until we are ready to use it (the neighborhood dog owners are thankful). It would take us much longer to add this element to our compost if we didn't have the osteria. We try to use everything we can in the kitchen. Although onion skins, carrot tops, and other veggie remnants get used in making stocks for soups, once cooked, those, too, get added to the compost bucket.

It's important to add the soil layer soon after the kitchen waste. The soil contains the microorganisms needed to speed up the decomposition process. This won't be the sweetest-smelling compost ever known—at least not yet. But that soil layer will really help cut down on the high "garbage" notes.

Just as in strictly biodynamic compost making, each layer must be lightly watered after being added in order to keep the material moist. Water is crucial in the proper heating and decomposition of all the layers. When the pile gets too dry, the aerobic microbial activity slows way down. We water the pile as necessary when watering the garden. Just as it's key to keep plants alive with water, the same holds true with the compost pile.

Conversely, too much water can be a problem as well. During rainy periods a lid, or a tarp, on the compost can be helpful. Heavy and consistent rain can wash away all the good nutrients you have been working so hard to foster.

There are all kinds of shapes and sizes of compost made in this fashion. Compost can be built in a pile above ground, or in a pit. Though, in a rainy area, the pit can often fill up with too much water, so aboveground is preferable. There are several different shapes and containers that can be made or bought: pallets, chicken wire, wooden boxes, enclosed composting bins that rotate. Or the compost can work simply in an open, mounded pile, or a windrow pile that is narrow and long.

We use primarily three different kinds of piles here at the home farm. We have a series of pallets that contain continually working compost, and we have open rounded piles. We also have a closed, rotating composter, which we tend to use when we have compost from one of our other piles that is close to being done; we then finish it in the enclosed container. But I would say our preference is to use the pallets and open piles because they provide excellent air circulation. The pallets are already naturally ventilated since there is space between the slats.

We've found the best times to prepare the compost piles are in the spring and in the fall when biological activity can be at its greatest. Too much heat or too much cold slows down the activity and can even kill the bacteria. But the compost can be kept going all year long, as we must with the waste we generate out of the kitchen.

Professional organic composters have found that preparing and working compost under a deciduous tree is beneficial. Oak trees are the best, as they

are for so many things related to soil activity. They are the best for mushrooms and truffles as well. But other deciduous trees can be good too, as long as they are not walnut or eucalyptus, which are toxic to many plants and organisms. Placing the pile about 6 feet or so away from the trunk of the tree discourages pests and insects close to the bark. The shade from the tree's canopy keeps the compost cool and moist when there is too much sun and wind. This little trick falls under that notion of forest-edge ecology, and we try our best to duplicate it.

We have compost under tree cover, and we have compost out in the open. Spatial considerations and the size of our compost have dictated where we can put our piles. With the hope of a new winery building in our near future, the composts under the tree will have to move. So we begin to think about how to adjust and plan compost for the future. We also have an idea to reshape and reorganize our winter garden, which would include a space for the composting projects. We think of building a shed roof for the compost, and we think about planting a tree, and in the end we think, why not do both?

Compost made in the classic French intensive way should be ready in about two to three months. It should look like soil and be sweet to the smell. One of the nice things about this method is that turning the pile is usually not necessary, or just one turn usually suffices. The object of the careful layering of types of material is to create natural aeration and a complete decomposition without the extra work of turning. But just like any element in the garden or on the farm, you must be attuned to what's happening. If the pile seems too dry or too wet, or not breaking down, add the material needed, and turn it. We try to pay attention to the compost just like we do our vegetable plants or our fruit trees and vines.

Varieties

When Caleb and I set out to plant our first eight raised beds in the walled garden, we knew we had specific needs and wants for our vegetables. At the time, local small farming was just starting to take off with small farmstands cropping up and CSA (community supported agriculture) farms happening in and outside of many smaller villages. The proliferation of farmer's markets that we have now hadn't sprouted yet, but we had two farmstands that had opened on the road that takes us to the osteria. We stopped religiously at both for meat and vegetables for the restaurant.

ONE FARMSTAND WAS OWNED by a retired farmer who just couldn't stop getting his hands dirty. He and his wife still owned a small parcel of land next to their old farm, which had been absorbed back into their larger farming family's property. They spend the summers up here in Vermont and the winters down south in Florida. Nick started out small, just a few beds here and there of the basics: tomatoes, herbs, zucchini, squash, potatoes, corn. But his offerings expanded every summer, as did the size of the beds. He built a more formal farmstand shed where his wife could offer her homemade pesto and pickles, and their son sold his dark amber maple syrup in old Jim Beam bourbon bottles (for no extra charge). We loved going to Nick and Theresa's to pick vegetables for the kitchen, our haul completely informing our menu. As we began to put our own land to work, we would ask Nick's advice on growing this or that, and he always shared generously of his knowledge. And even though we grow so much of our own produce now, we still support Nick's efforts and incorporate his specialties onto our menu.

For us, growing vegetables isn't just about the pleasure of growing something, the beauty of how a garden looks, or the beauty of the bounty itself, though these are large parts of why we do what we do. Growing for our menu at the osteria is of paramount importance to us. One of the reasons we started down this road of "grow your own" was that we wanted particular varieties or types of fruits and vegetables that other small farmers weren't growing in the area. With an old-fashioned Italian home kitchen menu focusing on heirloom recipes that we've collected over the last twenty years and more, we needed certain types of greens, onions, tomatoes, peppers, and herbs to re-create the flavors of the dishes we wanted to cook. We also felt very strongly that the flavors should be from here, grown in our soil, informed by our terroir. Vegetables, just like wine, respond to the local environment and growing conditions and have a terroir. In fact it is absurd to think that all the same facets wouldn't apply: the makeup of the soil, the microclimate, the geography of the landscape, the grower himself, and the variety—these all affect the expression and edible product of the plants. Just like vines and apples, some varieties work better in the landscape than others. We can grow so many things here in Vermont as annual plants, even Mediterranean artichokes that we "trick" into thinking they have overwintered. The exception,

of course, are plants that just can't survive our cold winters, like olive and citrus. With extra work and protection, though, we can even coax a fig tree to offer up rich and ripe brown and green fruit.

This idea about terroir and vegetables is not new, nor does it occur on a blank canvas. Every home gardener thinks about what he would like to cook on some level, and has to take into consideration what will grow where they live. Immigrant communities often find spaces where they can successfully plant and harvest the produce typical to their former homes. In Boston, there is a huge community garden supported by the Asian population in the city. Many of the plots are for families, but also many are for local restaurants. In London, Indian and Middle Eastern families find allotments where they can plant certain kinds of eggplant, peppers, and spices. In each of these places, the community may not be able to grow everything from their homeland that is familiar to them, but they can grow quite a lot.

In France and Italy, as can be imagined, the concept of terroir and vegetables are part of the patrimony. Now, through the auspices of organizations like Slow Food, certain varieties of plants and animals raised in certain areas have become protected. These ingredients are significant aspects of the local culture, and they are now beginning to be treated with the reverence they deserve, just as cultural and historical relevance is supported in architecture. Foods that marry a place, like a wine marries a landscape, are a crucial part of our history and heritage.

European chefs took up this mantle some time ago. We're a little slower here in the United States; for some time now our chefs have focused on locally raised produce and animals, but until quite recently we haven't had the same concerns with regional landscape and varietal preservation that the Europeans have had. Our food culture here is in its infancy compared with what is and has been happening in Europe. But we do see it beginning in areas like apple growing. We actually do have a history with tree fruit and cider; the apples that came to this country in the early colonial period are still being talked about and produced today. The Roxbury Russet is considered the best and first American culinary apple, and it happens to also be a cultivar excellent for cider. The Roxbury Russet was first discovered and named by Joseph Warren in Roxbury, Massachusetts, in the mid-seventeenth century. From Massachusetts it was then propagated in Connnecticut around 1649, and there is documentation that Thomas Jefferson planted them at Monticello in 1778.

Since Slow Food's development in this country, we too have programs that support the preservation of uniquely local American foods. Slow Food's Ark of Taste program not only travels in Europe, Africa, and Asia identifying and promoting small-scale food products that belong to a culture's history and tradition, but has an active involvement here in the United States as well. We actually do have traditions here in America that reach back into our native and tribal cultures as well as the early cultures that came to settle here. These cultures all contribute to our rich and vibrant food ways.

In Paris, the Michelin three-star chef Alain Passard of L'Arpege has chosen to highlight this idea of terroir and vegetables. In 2001, he began cooking exclusively *la cuisine legumiere*. We might consider this vegetarian cooking, but in Passard's philosophy it is much more than a simple label. By 2002, L'Arpege had its first potager to supply the restaurant. Passard bought a property, Chateau de Gros Chesnay, in the Sarthe region of Le Mans not too far from Paris. At the beginning, he was taken with the idea that he himself would garden there and bring the vegetables from the country into the city.

But with the size of the restaurant and the size of the garden that needs to feed it, he quickly understood that he could not do it alone. He hired a gardener, and eventually there would be five. Now, because of these ideas about terroir and vegetable growing, they found that some of the types of vegetables would do better in different kinds of soil, and they have grown beyond their initial 2-acre potager in Fillé sur Sarthe. A large parcel that faces the bay of Mont Saint Michel was planted, the potager Les Porteaux, as well as another in Normandy called Bois Giroult. Passard and his team of five gardeners found that the sandy soil of Fillé sur Sarthe grew excellent carrots, spring peas, bees, and honey. They produce most of their red vegetables and bulbous vegetables at Bois Giroult because the soil has more clay (red grapes often thrive in red clay-based soils) and is known to be beneficial for celeries, rutabagas, and potatoes. The third parcel, Les Porteaux, the one facing Mont Saint Michel, has rich, alluvial soil, providing a microclimate particularly good for aromatic plants. Everything is organic and horse-drawn and incorporates honeybees, hedges made from fruits, and the natural attributes of the landscape to aid in the health of the gardens.

Among the three potagers, Passard is producing 450 varieties of fruits and vegetables that get featured on the restaurant menu. The produce gets sent down on the TGV (the really fast train) every morning, and the compost from the restaurant goes back the next day, getting incorporated into the life of the garden. As might be expected, Passard's menu at L'Arpege corresponds heartily to the seasons.

We, too, respond to the seasonality of our garden, cooking according to the days and months of spring, summer, autumn, and winter. We choose our vegetables based on our library of dishes that we've recorded from Italian grandmothers and grandfatherly cooks who've allowed us in their kitchens and offered us sustenance at their table. This highly personal code of recipes comes from the inspiration sparked by the philosophies of these people and their cooking traditions. Our food is homey and homely; simple food that focuses on the purity of the ingredients. Now, for Caleb, the recipe has gone far beyond the memorized, written, spoken, or felt document or tradition. It begins with the seed planted in the ground. He begins cooking when he rakes the raised beds and carefully buries each seed or settles in each new transplant.

Our approach to the varieties we plant is manyfold. Firstly, we choose those specific varieties called for in our menus and ones that we know can grow here. There are several varieties in each category of plant cultivated. As an example, we plant about seven different kinds of tomatoes. We plant

them for different flavors and applications. We plant cherry tomatoes for pizza, we plant salad tomatoes, we plant for sauce. We plant them based on ripening time, so that we can have a steady stream of fruit from midsummer until the bitter and cold end of the growing season, or at least for a longer harvest period. And we plant them in profusion in case one kind doesn't take, but another does. The gardener has to take into consideration the odd packet of old, sterile seeds, or poor growing conditions for certain varieties to germinate. We plant heirlooms in order to keep the older true varieties that are well adapted to our climate naturally going. We do plant one hybrid— the Sungold cherry tomato—out of nostalgia for that first garden we shared together. Sungolds also happen to be very tasty in so many different ways.

In the walled garden and the raised beds that were built as a mirror image down below our house, the soil is fairly similar, so while we have begun to experiment with our terroir here on the home farm, these raised beds are more generalized because they are built up from compost and topsoil. We grow all kinds of things in these beds, everything from carrots to onions to bush and runner beans, to eggplant and wild arugula, escarole and radicchio. These roots and fruits are growing in soil with a beautifully aerated tilth. Having said that, below these mounds of compost and topsoil in the upper walled garden is clay, and in the lower gardens there is a significant amount of sand. In the lower beds we have set up an orchard paradigm with plum trees planted in the center of the raised beds of the potager, creating an environment conducive to plants like wild greens, roots, and herbs.

Our real efforts at finding interesting terroir for our vegetables is in the work we do out in the vineyard, and what we've slowly begun in the orchard. Just as in our two sections of potager, the vineyard has different soil components and different microclimates in certain sections of the field. The more we learn from our work with the vines, the more we learn about what we might plant in rows in between or beneath the vines or in areas still to be prepared for the planting of vines. Differences between clay, sand, moisture, forest edge, and longest periods of sun affect all these separate yet interworking blocks of vines and the vegetables planted among them.

We look to farmer's almanacs, classic horticulture texts, and what we learn from day to day and season to season and year to year on the farm to inform the ever-evolving plantings. It's so easy to misjudge where you think a plant might do well based on book-learned information. One season, we chose a spot for potatoes at the edge of the sandier side of the vineyard, knowing that potatoes prefer well-drained soils. We knew they liked soil rich in nitrogen, phosphorus, and potassium. This area next to the vineyard is slightly more acidic, also a plus for potato growing. This seemed like the perfect spot with potato-loving elements coming together. My oldest sister and niece joined us for a few days' visit, and all three of us trekked out to the far end of the vineyard and dug our trenches, burying the seed potatoes and backfilling them with soil. We were very proud, feeling like we were connecting to our Irish ancestry. But it turned out to only be a mediocre spot. There are so many things going on under the surface, and so many elements to consider, that it's not always a simple recipe to choose appropriately. Most of those plants didn't produce potatoes, though there were a few that produced really well. The green part of the plants succumbed to potato blight that year, the leaves just kind of shriveling and dying away. Blight, of course, proliferates in temperatures above 50 degrees Fahrenheit and high humidity. Rain can wash the blight spores into the soil, which then seep down and infect the tubers. We often have a perfect climate for the proliferation of the disease.

This is the same blight that starved out my father's family ancestors in Ireland and sent them to the New World. Scientists have spent scores of years trying to figure out crosses and varieties that will be more resistant, and while there are some that prove better adapted, the resistance is still quite low for most potatoes. It's interesting and not surprising to note that researchers have found that the wild species of potato, *Solanum verrucosum*, is able to readily resist late blight disease. Mother Nature always knows best.

What did we learn from that planting? Well, mostly that the soil was not thoroughly amended with compost. It drained too well, draining nutrients in addition to water. While I think that the soil had a pretty good nitrogen content, we suspect that the potassium in the soil was quite low due to other elements we noticed in the vineyard. Our weather that year was also very conducive to the growth of the blight inoculum. In subsequent years, we have planted potatoes in two new trial locations: on the edge of one of the large compost beds that has a rich mix of green, protein, and manure material, and in the potager raised beds themselves. The plants have thrived.

Seeds

The sources of seeds and how they are seeded can be one of the most important aspects of planting a garden. They are the raw materials of the garden recipe. We tend to do most of our seed buying from a company called Seeds from Italy, because we are very focused on particular vegetables that you can find

there. But we also buy from High Mowing (a biodynamic seed company based in Vermont), and Johnny's, a longtime source for many small organic farmers. We have also begun the adventure of our own seed saving, or of cultivating transplants that crop up in our compost pile from seeds discarded from vegetables we use at the restaurant. These are often our strongest specimens.

OUR PHILOSOPHY ABOUT SEEDING is a bit wild, like most other aspects of the farm. In the spring, Caleb's mother and her gardening partner start seeds for us in March in a heated nursery space they have. Primarily, they get our tomatoes going so we can get a jump on that season. We take a seasonal break twice a year from the osteria: in November and April. In the past, April was when we always tried to travel long distance, often back to Italy, which has inspired us so much, and we have tried many different spring seeding times around our April travel schedule. Now, going far away in April has become less possible given the pruning needs of the vineyards and more intensive seeding for the gardens. November, when the gardens, vineyard, and orchard have been put to bed, will become the time to really travel. Without the pressures of being away for an extended period, we can start seeds in April, and even March. Yet every spring is different, and since we rely on natural light and temperatures in our greenhouse, some early seedlings are successful and others aren't. We learn from trial and error. We try to seed onions, escaroles, chicories, and radicchios, any plants that like cooler weather for germination.

For those early spring starts, in the past we've used cells and potting soil amended with our compost. We've also seeded starts in just our own compost. These starts that we get going in just the compost seem to do much better in the richer nutrients, though commonly gardeners are not encouraged to use just compost to start seeds, as it is often a little too "hot" (too nutritious) for younger plants. We plan to continue experimenting with this ratio of the compost and a locally made potting soil until we formulate what works best.

SEED LIBRARY

Caleb is not the biggest fan of flats and cells, only because he finds that with our schedule and the schedule of the plants, it is not the best system for us. Ideally, once a seed has germinated and after about three weeks, the seedling should go into the ground. By midsummer, we have run out of plantable space and often don't have time to completely prepare a new bed out in the field once everything is all going at once. So sometimes our starts will linger in their cells, and as a consequence they are not strong plants.

For our needs, we've found that in a some ways once the season really gets going we much prefer broadcast seeding. We've found that when we can prepare raised beds out in the field, and broadcast seed on them throughout the season for a succession of crops, these plants are much stronger and happier. Those later crops that are destined to go into the winter garden have a much better chance at being productive and happy plants if they've started out in the field rather than the cell, even if they've been of the better specimens from the cell.

The reasons are probably fairly obvious: When a plant has more room to stretch and grow, it will stretch and grow. We've found the root structure of the broadcast-seeded plants is much stronger, especially the tiny, fine root systems that we can't even see well but that do most of the heavy lifting of micronutrients. The plants are able to become established, and while it is always a shock for a plant to be dug up from its home and transplanted, these plants always rally.

This coming year, we will continue to have Carol, Caleb's mom, get those tomatoes going inside, and now that we have a little more space devoted to plants inside our own house, we will get some of the onions, escarole, and chicories going. We will also seed cool-climate-loving plants in flats in the winter garden if the weather cooperates and the winter temperatures begin to rise and the sun starts to shine more often.

Every gardener finds his or her own tricks and preferences through trial and error. A garden must fit into your life in a way that dovetails with your schedule. If you can't find the time to get those midseason veggie starts in their proper bed and you watch them languish in their cells, you change your approach. There are definitely set precepts and examples of how to go about gardening, and these are invaluable in helping get started. But in the end, every gardener's garden is individual and idiosyncratic, and is connected not only with the life of the gardener, but with the gardener herself.

Seed Saving

I've heard it said that to be a seed saver you help change the world we live in. I like to think of it more in terms of preserving the world we live in. And of harking back to a world we used to live in.

SEED SAVERS ARE a particular brand of people. Their gardens can be inherently more wild looking as they let plants mature past harvest, and to some very regimented gardeners it would appear a seed saver's space is unkempt. But seed savers are not gardeners with a devil-may-care attitude. They must work precisely, taking the plant to that ultimate maturity in order to harvest the seed. A good friend of ours who is a seed-saving farmer has beautifully maintained gardens, but her rows and beds have a different aspect because of the goals of the gardener. The colors are different, more saturated and darker because, I imagine, the plants are going past the optimum time for food harvesting when the colors go from bright jewel tones and begin to darken and fade toward overripeness. It reminds me of the colors in the month of August; everything turns a little darker and more golden; the efforts of photosynthesis shift. The plants mature and age.

Seed saving is an art different from the art of growing food. The seed saver looks ahead to future harvests; aside from any food growing they do outside of their seed saving, they don't think about harvesting food at all, but the *potential* for food. They look back, they harvest, and they harness the energy and strength of the variety in order to safeguard its longevity.

I think most serious gardeners and farmers come to a crossroads at some point in their journeys where they are attracted to the possibilities of seed saving and are smitten by the idea of sustaining their own parcel with seed generated on the farm and in the garden, seed that they have brought full circle through life and harvest, a neat algorithm of the calculus of our lives.

CARROT PLANTS IN FLOWER

This is the way farms used to operate: Each farm collected and saved seeds to plant for the next year. With the serious industrialization of agriculture following both world wars, many farmers and home gardeners began to purchase seeds from commercial seed suppliers, especially as hybrids became popular ways to increase yields. Seed saving now is very much a grassroots movement, mostly found in the home gardener realm. Though more and more midsized farms interested in organic and ecologically sound methods are starting to save seeds, too.

There are many reasons that seed saving is a significant aspect of cultivating the land in our current environment. Plants that reproduce through nature tend to adapt to the local terroir over time and evolve as dependable varieties. "Folk" or heirloom varieties grow in connection with their landscape, and historically they were often referred to as *landraces* by farmers and horticulturists. A true landrace was often christened with a local name by local farmers and was a native variety that had particular characteristics that had adapted to local climatic conditions, cultural practices, and diseases and pests. They were genetically diverse and able to respond to seasonal changes. With the advent and increased popularity of hybrid varieties, where scientists try to crossbreed plants to specific requirements, usually for increased yield and uniformity in size for ease of mechanized harvesting, landraces became at risk of extinction. In the last sixty or more years, landrace or traditional varieties of thousands of vegetables and flowers have been lost because of this increased reliance on commercial hybrid seed.

As in so many things, there has been that swing back to older ways of cultivation. Concerns about the erosion of the horticultural gene pool, which results in less-hardy plants that are more vulnerable to diseases, pests, and changes in local microclimates, have been one of the primary motivators for seed savers. Traditional varieties have open pollination, or create plants that reproduce naturally from saved seed and adapt to local conditions. Hybridized plants tend to negate these natural evolutionary elements and cannot reproduce themselves authentically or genuinely. The move back toward more genetically diverse seeds and recapturing our heirloom horticultural history has sparked a pronounced discussion and effort.

Seed saving is not just about collecting seeds from harvested plants or allowing plants to go to seed and then collecting. For seed saving to be successful, the farmer must develop a new set of skills because, even though you are allowing the plants to evolve naturally, you still need to be aware of the desired and best characteristic of the specific landrace. You need to learn things like how plants of each variety must be grown together in a way that will preserve their inherent genetic diversity. Tomatoes are self-pollinating, and if you avoid planting hybrids, you can grow the same landrace from year to year, even if other tomato varieties are grown relatively close by. This is not the case with plants like peppers and eggplants. Their flowers can be cross-pollinated by insects, so different varieties of these must be separated by at least 500 feet for the seeds to remain pure. Squash, cucumbers, and melons need even more space, as they are all pollinated by insects. Close varietal kin need to be separated by ½ mile or more in the case of these sprawling cucurbits if you want to maintain the original seed strain.

Proper seed storage to maintain the viability of the seed is paramount in learning to save, as is learning the situations that will maximize the seed's ability to germinate. These are just a few of the intricacies crucial to embarking on a seed-saving adventure.

For the beginner, vegetables that have seeds that are easy to collect and process after harvesting are a great way to start. Cucumbers, tomatoes, melons, zucchini, squash, beans, and peas are all relatively easy crops with which to learn. In the natural way of things in the garden, if you didn't pick all your cucumbers, they would eventually fall to the ground, rot, and lay seed in the soil that would most likely germinate the next spring when the soil was warm enough. When you save seeds from an overripe, yellowing cucumber in the garden, you are just managing that natural process and harvesting the seeds once they are fully mature, but before they try to plant themselves.

BADALUCCO BEAN PODS

Our own efforts at seed saving have just begun. Just like any beginner, we started off with the easier seeds. Our first trial was with pole and bush beans. We let the pods dry brown before harvesting for seeds, which is about six weeks after the eating stage. We knew to pull the entire plants, including the roots, if frost threatened, then to hang them in a cool, dry location until the pods had turned completely brown. But we didn't have any frost problems that first year.

Beans are easy to process for seeds, and because we are a small farm we could do it all by hand while sitting around before dinner. We opened the pods and removed any chaff and winnowed any remaining dross. We stored the shiny cream-colored, black-speckled beans in the cool dark part of our pantry and planted them the following spring. We have been propagating beans from our own bean seed for two years now. Little by little we add another seed to learn about, and try segregating some plants from the crop that we don't harvest for the vegetables. We are fortunate to have that good friend who lives not too far away whose specialty is seed saving. We've learned an incredible amount from her gardens, her seeds, and her tutelage.

SIMPLE SEED SAVING

The easiest seeds to save are those from peppers and melons. The seeds are mature when the fruits have changed into their final color. Separate the seeds from the flesh and let the seeds dry in a non-humid, shaded place. All wet seeds should be dried on a glass or ceramic plate. Keep them spread out and not clumped together. Don't dry your seeds on paper towel or a paper plate as they will stick to it and you won't be able to separate them from the paper. You can use a food dehydrator set at 85 degrees Fahrenheit, but don't dry them in a warm oven or anyplace where it's hotter than 95 degrees Fahrenheit.

Tomatoes are easy, too, but they take a bit more time. Tomato seeds are covered in a pulpy, gelatinous coat—this is actually what prevents the seeds from sprouting inside the tomato. To remove this coating, you have to ferment the seeds first, which imitates the natural processes of the rotting tomato and also has the added advantage of killing off any seedborne diseases that might affect your tomato crop next year.

To ferment the tomato seeds, add about half the amount of water as there are tomato seeds and juice in the bowl. Stir the mixture twice a day for about three days. This is not unlike how you would handle newly fermenting wine in the cellar. Be aware of what's going on in your bowl. If it's warm out, fermentation will begin more quickly. As the mixture ferments, the surface of the liquid will become covered with white or gray mold. Note to yourself to keep your fermenting bowl in a place where it won't be disturbed or smelled. It can get pretty stinky.

Once a thick film of mold has appeared or the liquid has started to bubble, you're ready to get your seeds. Stop the fermentation by adding enough water to double the mixture, and stir vigorously. The clean, viable seeds will settle to the bottom of the bowl. Gently pour off the mold, liquid, and floating seeds. Add more water, and repeat the process until only clean seeds remain.

To capture the seeds, pour them and the remaining liquid through a strainer. A little trick to help them dry is to pat the bottom of the strainer with a towel to remove as much water as possible before dumping them out onto a glass or ceramic plate to dry. Mix them twice a day to ensure even drying and prevent clumping. Tomato seeds need to dry quickly, as they will sprout if left with

moisture. A gentle fan can be used, but don't put the seeds in sunlight, or in that warm oven, no matter how impatient you get.

There are slight variations for seed saving for all vegetables, but these are at least two simple ways to get started.

A great resource to learn more about seed saving at any level is the Seed Savers Exchange, which began in 1975 as a response to big agriculture and the loss of biodiversity in the garden. The goal was to offer an alternative model, to conserve and promote our culturally diverse garden and food crops, and to encourage participation with collecting, growing, and sharing heirloom seeds and plants.

The walled garden started with eight beds on that upper terrace just above our farmhouse. Very soon, that wasn't enough. We needed more room to grow: to grow our ideas and all the produce and flora that went with them. Our house is built into a slight incline, and at the back of the house both the main floor and the bottom floor are exposed. On the main floor is a wide balcony and on the ground floor is our bedroom and office, one large room the width of the house that opens out onto a terrace with a rose- and grape-vine-clad pergola. Off the terrace is a wide, relatively flat space, the same space where we pitched that grand tent and we served our friend's wedding dinner in the driving rain. Just a year later, that space mirrored the original walled garden. We had another eight raised beds with gravel paths and four plum trees anchoring the space, a rustic wooden fence encircling the end of the apron of paths. Another set of stone walls embraced the whole of the terrace and garden beds.

THIS WALLED GARDEN IS FRAMED by a hedge of red, antiqued roses, and the stone walls are covered in classic, pink William Baffin climbers and topped with lavender, hydrangea, and Korean lilac. The beds themselves are constantly in rotation, but in orchard fashion we concentrate on escarole, chicories, and wild dandelion greens, onions, carrots, and bush beans beneath the fruit trees. A pole bean trellis went up, and every year it gets

moved to a different empty bed, creating a wall of cascading green leaves and pods. Blue-flowered borage and winter pansies have seeded themselves in the beds, mixing into the rose hedge and also at the corners of the beds, softening the hard lines of the regimented boxes.

In the upper walled garden, guests come to sit and taste wines at the long table in the center. On any given night off from the osteria in the summer, after we come in from the field, we will adjourn to the table lighted with candles for an aperitivo to salute the end of the day; we pretend we are guests ourselves, escaped to somewhere in the country to a little place that has a lush garden where they serve glasses of chilled sparkling rosé and plates of little treats to be eaten as the sunlight fades and a broad moon rises.

In the lower walled garden, the terrace has become the stage set for long, vibrant-tasting lunches and dinners. That stage has seen much laughter, friendly debate, and lively, mercurial discourse. At that table, we've served

dishes embroidered with the flavors of our land and our neighbors' land. We've tasted wines paired with vintage perfumes inspired by places so very far from here. We've watched short films on a sheet hanging at the end of the balcony, seen and heard a celluloid wolf howl at the red swirl of his hooded mistress, and as the full Buck Moon rose over the hill to the east, a pack of coyotes at the bottom of the field, at the edge of forest, answered his cinematic call in full-throated voice. All these things offer certain magic.

The walled garden, like the house at the epicenter of the garden's influence, like the table that sits in each stone embrace, is at the heart of the farm. It feeds us, feeds our friends, and feeds the osteria; it feeds our compost, and in turn feeds the vineyard, the orchard, and the garden, once again offering sustenance and courage for the season to come.

Winter Garden

The winter garden is an extension of our walled garden, the proverbial lion in winter, the strong, kingly creature confronting adversity and age in the chill of December, the snows of January, the freeze of February, and the slow, raw awakening of March. In like a lion, out like a lamb. But for us, in northern New England, April can still present the rangy wildcat. Caleb's grandmother always used to say that Vermont spring was the last four snowstorms in April.

THE WINTER GARDEN, the pared-down form of our summer bounty, has to face the challenge of deeply cold temperatures, waning sunlight, and often very cloudy skies with just a pale wash of light that must inspire the greens planted to continue despite the threats of our alpine season. There are many

winter greenhouses in our part of the world that work with heat or partial heat: woodstoves, compost-driven furnaces, electricity, and gas. But our hope was to plant a greenhouse that would heat passively and wouldn't need additional fuel to keep going through our shortest and coldest days.

In 1998, the organic market farmer and educator Eliot Coleman and his gardening author wife, Barbara Damrosch, thumbed their noses at the months of October through May in defiance of the long, cold, and often bleak winters of Maine where they live by developing an environmentally sound, energy-efficient, and viable way to garden through these brooding months and to grow a winter harvest.

Through research and practice, they created a minimalistic approach to the non-tradition of New England winter farming by focusing on planting cold-hardy varieties grown in and with simple greenhouse forms and methods that are appropriate to the needs of these kinds of crops. Their philosophy was that the cold-hardy crops were correct not only for the season but also for our menus, our tables, and, by natural extension, our bodies; that the essential design using high tunnel greenhouses would offer new prospects for farmers and home gardeners in any part of our country where winter weather puts a damper on the longevity of the growing season. Their theories

THE WINTER GREENHOUSE

of winter farming and harvesting have spurred a whole generation of young and older gardeners alike to embrace our winter months and grow what we can by using the resources that we have, along with a little ingenuity.

Our area of Vermont is on the same latitude as the Renaissance city of Florence in Italy. All of Italy has centuries-old, if not millennia-long, traditions of growing winter vegetables. We have the same amount of daylight during our winter months. Certainly, we have colder temperatures; for a variety of reasons their climate is more moderate. But with the clever construction of a hoophouse and raised beds covered with an extra layer of row plastic, we can mimic enough of the growing conditions they have in that part of the world, their climate supported by the warming influence of the Gulf Stream.

There are several components to the effective winter garden: the greenhouse, the row covers, the cold-hardy vegetables, and the use of succession planting. Our greenhouse is a traditional hoophouse, or high tunnel with arced metal ribs and metal spine. We came by this structure through sheer luck. Friends of ours who farm in the village next to ours, and who make a stunning alpine cheese, had bought a hoophouse several years before, thinking they would plant it for winter gardening for their own kitchen. But the demands of a dairy farm and a cheese house kept the building of the hoop-house always at the end of their list of chores. When they heard we were on the hunt for a frame with which to get started, they kindly offered theirs to us, as the structure was just sitting on their farm, unassembled and taking up space. They were happy for us to put it to use; in trade they would be welcome to winter greens whenever they wanted.

Our inherited curved hoops and spine are covered with UV-resistant plastic with wooden end walls that have doors for airflow during warmer times of year. Our growing space started out as 20 feet by 24 feet, but it has grown incrementally over the last few years to 32 feet in length. Not only have we realized that we need more and more winter growing space, but the theory is that the more things you have growing in your winter garden, the warmer the whole will be. All those little plants take in carbon dioxide and release oxygen, which creates a certain natural greenhouse effect.

The raised beds themselves are 10 feet by 4 feet and canted on an angle so that one end is higher than the other, tilting the beds toward the side of the greenhouse that receives the most light during the winter, responding to the trajectory of the sun at that time of year. They are built out of hemlock, but you can use any good, untreated wood that is rot-resistant.

Almost any lightweight translucent fabrics that allow air and water to pass into the beds are really good for row covers. In the past we've used old windows, taken out of our local library when it was being renovated, and we often use pieces of the same material that covers the hoophouse itself. Frankly, neither of these covers are ideal, as they don't allow water to seep in, but they do let light in. We had both materials on hand, so they were readily available at no extra expense.

The whole concept of winter gardening involves being savvy and using what you have. Because we've used windows and the UV-resistant plastic for the row covers, we had to adapt them to allow air and moisture into the beds. For the windows, they were secured with hinges on one end, the higher end of the beds, and simply propped open during the warmest hours of the day for air circulation. With the plastic, we have anchored one end of the mate-rial to the bottom of the beds, and then stapled strips of wood to the opposite ends on the higher side of the beds. The plastic is cut exactly to the width of the beds and the unsecured end is cut longer and drapes over the back, weighted down by the length of narrow wood. During the warmer parts of sunny winter days, we roll the plastic back around the strip of wood and the end of the wood rests on the sides of the boxes, keeping the plastic up and away from the plants.

Sometimes, moisture collects on the top of the plastic—this is when a permeable row cover would really be better, as it takes in the water naturally. These puddles of water can weigh down the plastic and crush the plants, creating additional condensation in the bed. Condensation can be a good water source for the plants, but when the plastic is pressed against the leaves, it can create rotting. Work with what you've got as a remedy. We use plant pots set upside down and slightly higher than the beds to prop the plastic up and away from the plants.

These floating row covers act like an additional greenhouse, tempering the microclimate beneath them. The raised beds have extra protection, which creates more humidity in the world beneath the row covers. Additional humidity and moisture in the beds protects against the freezing temperatures. It's just like in the vineyards in Champagne or the cranberry bogs on the North Shore of Massachusetts, where growers will continually douse the plants and crops with water during the night when there is threat of freezing. The water protects the plants themselves from freezing and prevents the loss of the fruit. The same principle is at work here.

In our winter greenhouse environment, below-freezing temperatures still occur, but the temperatures don't get as low and they don't stress the plants in the same way as if they were outside in the elements of the season. This whole idea of the inner and outer layer of covering is the basis of how this low-energy winter garden works. Eliot Coleman describes it quite eloquently: He says it's like the layering of our own warming winter clothes, "like wearing a sweater under a windbreaker."

The planting of cold-hardy vegetables is also key to the whole concept. It would be foolhardy to try to grow frost-tender crops like eggplant or tomatoes at this time of year. But there are plenty of plants that prefer cooler temperatures and like to defy the idea of winter. Spinach, chard, carrots, scallions, arugula, and buckhorn are all good standards. But we've found that herbs like parsley, oregano, rosemary, and sage can continue to thrive in these circumstances, too. Bitter greens love cooler climes: Chicory, dandelion,

endive, escarole, kale, radicchio, and collards are often quite happy in the winter garden. Radishes also tend to thrive. These cold-hardy vegetables are even at their best in the cooler temperatures of fall, winter, and spring; their flavors coalesce and burst forth with a bright, tender intensity that is shut down in the heat of summer.

We've concentrated our own efforts on the greens in the winter greenhouse because of our current amount of space (which is never enough), leaving the green onions, leeks, and root vegetables like carrots to live protected in the outdoor garden under the blanket of snow, which tends to sweeten them. We dig the more fragile onions out before there is too much snow cover, and we spend the time chiseling out the carrots from the soil in the middle of the winter's tale. We also leave several carrots in the ground to await an early spring harvest once the snows are gone. These are the sweetest of all.

One of the questions often asked at the osteria about the winter garden is: How can we keep things growing in such cold temperatures? During the coldest and darkest part of the winter, the month of January through the middle of February, the greens in the greenhouse aren't growing; they are holding steady. We've realized that if we put enough plants in the ground in November and December (depending on how mild the season is) we can harvest leaves all the way up until the light shifts in mid-February. Plants begin to grow again after that time. But if we are short on space and plantings, we might get through mid-January, but see no return of leaf growth until that magical moment in February. As a consequence, our greenhouse has grown just a little bit each year, with new plans for yet another expansion to accommodate even more.

The key to making this kind of winter gardening work is not just having the right plant varieties and the correct infrastructure for growing; instead it's really reliant on the concept of succession planting. Seeding for a winter garden starts in August, which seems completely backward when you have been gardening according to the regular seasonal patterns. Because you have to reverse your thinking about how and when to get things going, it can take some time to get into the rhythm of a winter growing season. The length of the day has already begun to change by August when you need to get those first seeds into soil, so the decreasing daylight hours as well as the cooling of temperatures have to be taken into account. The timing of the seed sowing is very crucial because of these factors. The goal is to have relatively mature plants to put into the greenhouse. Winter-hardy vegetables fortunately germinate better at slightly cooler temperatures, but they do still need a certain amount of light each day.

Plants that have been growing outside during the height of summer will be less hardy during the winter months if transplanted. Plants that are started later in the summer and are transplanted into the winter greenhouse will do much better, as they are more acclimated to the cooler air and decreased light. We have also found that the seeds sown directly in the ground out in the field later in the summer that are then moved to the greenhouse in the late fall and early winter (if the ground is still soft) fare even that much better. Those started in cells, even with some of our own compost, just aren't quite as robust during the most brutal days of winter.

SPRING STARTS IN THE GREENHOUSE

Our winter greenhouse has ended up having a dual and unexpected purpose: It has become a year-round working space for our plants. In the spring, many of those cold-hardy plants come back outside to the raised beds to produce for spring and summer months, or as long they will produce, and tomato starts and other early transplants go in. In the process of transfer, the soil gets reworked, the sides of the greenhouse get rolled up to let air travel through, and we plant vegetables like tomatoes, peppers, eggplant, and basil under the cover of the greenhouse. And there they stay for the season. Because these are all plants that love consistent heat, we can provide that for them just by harnessing the effects of the sun in the greenhouse space. We can manage the light and heat with the rolling sides and a shadecloth that drapes over the top.

A couple of years ago, our area was affected by a serious tomato blight. Farmer friends dependent on these fruits were sorely impacted by the loss of their tomato crop. Timing—or rather a lack of enough time to do all the chores we need to do on a given day—plus a little luck, and a suspicion at the back of our minds, a tiny intuition that it was a good idea to keep our toma-

THE WINTER GREENHOUSE IN SUMMER

toes in the greenhouse that summer, proved very beneficial. Before that we used to plant all our tomatoes outside; some years were good, other were not. When we kept the tomatoes in this protected environment, not only did we not get the blight, which was airborne from other properties, but we extended our tomato growing season by at least a month. Because of this lucky accident we began to think that planting the tomatoes and other heat-loving vegetables in the greenhouse for the summer could be a smart direction.

Our tomato growing power has increased exponentially since incorporating the greenhouse into our summer growing, and we've been able to incorporate succession planting for our tomatoes to last well into the fall. Even if we have tomatoes still ripening when it starts to get too cold for the tomatoes outside, we harvest them and bring them indoors in our grape-picking baskets to finish turning color. We often add an apple to each basket, as pome fruits help tomatoes ripen because they add to the ethylene gas that tomatoes produce and speed up the ripening process. In this way, we've been able offer our own harvested tomatoes at the restaurant for much longer than the traditional season, with no extra energy output. This past year, we served our own tomatoes in special pasta dishes and on our thin-crusted pizza until Christmas.

I love the happy accident, and I love when one small step leads you to another story, another idea. In our efforts to keep the winter garden going, not only do we have the greenhouse up and running, but we have begun to make space indoors for more delicate herbs and plants. As an experiment, we brought in a mature eggplant plant that is quite healthy and strong. It sits near a French door where it gets light but is not too warm. If it makes it through the winter, we'll plant it when the ground is warm enough and see if it begins to put out flowers and pollinate.

Because of this not-so-little indoor eggplant, we recently heard about a gardener here in Vermont who brings her tomato plants in at the end of the season. The vines harden off just like grape vines, and she transplants them again at the beginning of the summer to produce fruit. She has gone several seasons now with perennial tomato plants. An inspiring idea—one we already have slated to try for next year.

The Winter Greenhouse as been such a positive addition to the farm, and in so many ways. The obvious benefits are the ability to extend the growing season through our cold months and to harvest so much for ourselves and the osteria. But there is another side to all this productivity for the table and the larder; there is the sheer balm of working and being in a space in which green things are growing when the air is cold outside and the ground is covered in snow.

ON ANY GIVEN WINTER DAY, we bundle up and trundle out to the winter garden. The glass-paned doors will be covered with ice crystals that look like a filigreed glass etching of snowflakes. Or the glass will be wet and slightly foggy because of the heat being generated inside. On those days, the jackets come off quickly and hang on hooks on the wooden wall once we are inside and working. The radio goes to the local classical station; it's true that plants respond to harmonic music. At the back of the greenhouse sits a table with a few café chairs and an old oilcloth covered with red and white checks and a pattern of fruit baskets. A place to stop for a cup of tea brought out in the big red thermos painted with blooming pink camellias. Or to settle in for a winter lunch on a sunny day when the thermometer rises. We'll bring out a picnic of sausages and onions or even a pasta decorated in fresh herbs with little glasses of our own red wine. Our neighbor will come and join us. On those sunny days, you can feel the warming rays through the

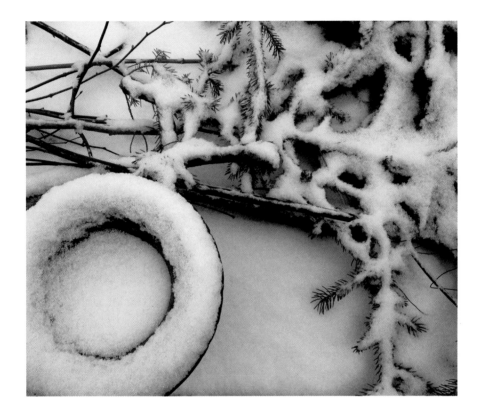

roof and walls of the translucent greenhouse going straight into the center of your bones. The air will be slightly humid, with the smell of black dirt and green leaves. It is intoxicating. Outside, the ground is covered by diamond-sparkled snow. The vines to the west of the greenhouse are just silhouettes of themselves, dark brown stiffly curling wood, messy in their pre-pruned state. The rose garden directly in front of the greenhouse comprises beds filled with dried parchment flowers of hydrangea and forgotten rose hips left for foraging birds. A few frozen and rotting apples remain on branches in the orchard; again, more fodder for birds, who will get slightly drunk, but warm, on the fermenting fruit. In the midst of all the grays, stark white, browns, and faded, tea-tinted reds of winter hibernation and decay, we have this small kernel of green, a composed world made of scent, vibrant color, and possibility.

Rose Garden

Gardens are intensely personal. They are an expression of the gardener, that fifth facet of what is called terroir. Some gardens are started from scratching at fallow dirt, some gardens are rehabilitated from overgrown plants, some gardens are rebuilt because they are not to the gardener's style, some gardens are inherited. Some gardens are extremely ordered, some gardens are innately wild, some gardens are unkempt, some gardens are forgotten, some gardens are constantly worked and lived in. Some gardens are remembered.

THE FLOWER GARDENS here at our home farm are a little bit of all of the above. We moved into the house on the property in the snowy chill of January

and February, and we didn't really remember anything about the plantings around the house from the three times we came to look before making our offer. We were so focused on the house and the view and just the fact that there was some land. I lie a little: We did vaguely remember something about an asparagus patch behind the house, but other than that we didn't recall flower beds or vegetables. And while we didn't remember specific plantings, we were made aware of one.

When we signed the contract in late December, there was a caveat tucked neatly in the pages. A large hosta plant that sat at the top of the property somewhere near the road was meant for a neighbor and not to be included in the sale. I couldn't remember this particular hosta, but because of some disagreements that cropped up over the sale with the previous owners, this exclusion felt a bit mean and sadly has forever colored my relationship with hostas. The following summer, when the classic white-and green-striped hosta made itself known and the neighbor down the street came somewhat reluctantly to dig the plant up and take it with her, I was frankly glad to see it go, and, as a consequence, I have never been a particular fan of the variety.

When that first spring did arrive and a few tendrils of asparagus poked through the ground, we found an odd assortment of flowers growing in unlikely places. A purplish-pink azalea planted in the middle of the north side yard stood alone and rather desperate looking off to the side. Nothing else surrounded it. In April, a strange and very straight line of daffodils cropped up in the same side yard. While they looked stranded between our lawn and the meadow adjacent, they made pretty and fragrant bouquets for the house. The assortment of them was also very wide: traditional yellow daffs, delicate and orange-centered jonquils, white narcissus with a delicate red rim around the pistil, and short double blossoms of yellow with spring green stripes.

A few other gems were found along the stone wall that separated our house from our meadow at the time. Two varieties of early June-blooming roses, thick mats of waxy-leaved vinca, a hearty clump of Persian coneflower, a few stray pink and white foxglove. There was no apparent rhyme or reason to the plantings and it was unclear who had put them there.

When we asked a neighbor whose family had lived in the area for the past five generations, he said that the original owner had had a fantastic wildflower garden that had run along a split-rail fence on the north side of the house. She would transplant wildflowers from the roadside and occasional cultivated specimens as well. The lone azalea had been at the head of the fence, and the daffs had been a part of her collection. The roses on the

stone wall were wild and hailed from the seaside from which she originally came, but he didn't know the varieties. One was a very pretty old-fashioned five-petaled classic with a sunny yellow center, almost like a dogwood flower. The second reminded me of a Tudor rose, white with a brush of pink near the heart of the flower that intensified as it unfurled and matured, becoming almost dusky.

It appeared that we had inherited a garden. Like the little house in the orchard and the wild, walled garden, an inherited garden was another thing of desire. Like the gardens that Thomas Jefferson had inherited in Shadwell from his parents, or the mysterious and exotic garden in an essay I had read by a woman who had inherited her mother's garden in New Orleans; like my own aunt, who bought and moved into my grandmother's house after my grandmother had passed on and inherited the garden I remembered from family gatherings and photographs.

I like the idea of the continuation of garden from one gardener to the next; a line of history is reaped in the petals and leaves of a diverse and intricately layered flora much like a house inherited is a collage of the layered personalities of those who have lived there. The inherited garden is so compelling in its stratum of terroir and the hands and spirits of the growers. As in vineyards, which are handed down from generation to generation, each

SIBERIAN IRISES AND WILD HONEYSUCKLE

winegrower adds his or her soul to the landscape and the fruit. Now, I too had inherited the remnants of a garden from a woman I had never met and never will meet.

No one could tell me much about my benefactor other than that her absolute pride was her wildflower garden that marched down the field along that fence, and that she had moved here to our alpine geography from somewhere near the sea—a young bride coming from the coast to live on her husband's family farm that had once encompassed our land. From her came the asparagus bed and the little potting shed for the lambs.

One set of owners separated me from her, and in that time the results of her labors had been dismantled by different desires, neglect, and mowing. Gone was the split-rail fence; gone was the fantastic area of local and imported blossoming plants. But nature is so resilient, and in that resilience also lies the spirit of a gardener. Seeds from those plants spread elsewhere, and even now, so many years after we moved to this place, new wildflower specimens will crop up unexpectedly in odd corners of the property. I see this strength of character present in the utter defiance of the row of daffodils that appear every spring. And every spring, I move them after blooming. I've planted them in proper beds. Every spring, a new crop comes up in the same location spread out along the phantom fence line, and I can't imagine how they managed to hide from my spade the year before. And each spring, though I'm so sure that I have certainly removed them all, they laugh mischievously with their flouncing bonnets, firmly entrenched.

When we broke ground on the original potager in the meadow, the flowers were kept closer to the house. I toiled, and I will freely say toiled here, as cleaning out the rose bed along the stone wall that originally separated us from the potager, mired in the vinca ground cover, was a fierce and exhausting job with only momentary gratification. I knew from my own mother how to weed and cut back, how to tidy flower beds, but the reality was that it had been too long for this bed of roses that was no bed of roses. Or perhaps it really did encapsulate the truer meaning of a bed of roses because a bed of roses is pretty to the eye and nose with its seductive scent and sweet aspect, but roses have thorns, and thorns are in no way comfortable to lie upon. They can prick you with surprise. They can draw blood. This was not a comfortable gardening project, and the hard-won upkeep was ephemeral.

The remnants of my lady gardener's work did not a garden make, and as herbs and greens grew in the potager, we made an attempt to build flower beds cradling the house. An ill-fated attempt if ever there was one. Clearing the space for the beds was all done by hand in ground that hadn't seen a fork or a hoe in a very long time. Our mothers brought us plants to fill these beds: feverfew, golden glow, mints, yarrow, tansy, scented geranium. A good friend with a fantastic garden gave us a variety of iris and loosestrife. Inheritances in another way.

What Not to Do

All these first beds were always too small in scale and didn't fit the location. I would set up a bed, plant it with all these green gifts or curious specimens that I would find while dawdling at the local plant nursery. Then I would move on to start another bed elsewhere, especially since I knew in my bones that the beds I was creating weren't quite right. I was trying to understand the marriage between cultivation and nature; I was trying to create something that blended into our landscape, that helped our farmhouse, this human structure, blend into the slight and natural decline of our hill. Subconsciously, I am sure that I figured that my next effort would be better, that I would have learned something. This went on for a couple of years, and as the numbers of flower beds grew, so did my frustration. I couldn't properly keep up with what I was creating, and I had also planted these beds with highly invasive species, as I was eager to have full-looking, mature, and slightly wild English-style gardens. The wild was no problem. In fact, the gardens took on more than a wild look; they appeared almost abandoned and feral. Mint and tansy took over like wildfire, crowding out other, more delicate models. My mother said she had warned me to not plant things like mint and tansy in beds shared with others. I felt I was doing a disservice to the memory of my lady benefactor. I was no better at bringing back her hopeful gardens than someone who had let them go completely.

IT WAS A DARK PERIOD for gardens at the top of our hill.

As we began to do more serious work on the house and my initial work at hardscapes snug up against the walls had to give way to the carpenters' needs of getting in and out of the house through various doors and windows and tromping through once imagined flower beds, those hard-won but misplaced beds gone in a nonce, I set my sights on our languishing potager. We had started that garden when we were running a bakery and a lunch and dinner spot, all out of the same kitchen and same dining room and shop. There was so little time for gardening. But when we realized it was time to retire the bakery, and a year after that retire our lunch service, suddenly our day spread out in front of us: A wide, open space beckoned before we had to go into the osteria to prepare for the dinner hour.

We attacked the potager first. And it was an attack in the sheer energy and perseverance it took to hew this space. It was a restart, a re-envisioning. The four parallel beds became a more traditional and monastic culinary garden shape: four square beds within a square. We plowed, we tilled, we drew lines in the dirt and used string to mark off spaces. Black landscape cloth went down and a load of pretty pearl-shaped variegated pea gravel went on top. I began to fork and hoe and plant. In went the dug-up roses from our lady gardener, in went rugosa roses from the nursery, then two bridal wreathes from the gardens where Caleb grew up. In went a quantity of Siberian and bearded irises from our good friend the gardener. In went two

clumps of heavily scented Hyperion lilies that were once in Caleb's grand-father's garden, as well as some wild clumps of orange lilies found on the roadside of our hill. In went tansy and mint from my mother. (I still didn't heed her advice to plant tansy and mint far away from any kind of formalized garden.) I wanted voluptuous gardens that looked like they had been here forever, not slow-growing plants that looked "new" for a couple of years. I already had in my head that wild walled garden near my parents. And while this garden didn't have any kind of structure around it, I wanted it to be as naturalistic as possible. I have come to realize that, as I continue to create and re-create gardens here at the farm, I am always trying to duplicate some-thing of that wild walled garden that continues to inspire, to haunt me.

Later that summer and into the following year we began to build the boxes around the beds. Somehow we missed that you design the beds, then you build the boxes, then you build up the soil in the beds to raise them. Only after you have done all that do you plant. We did everything back-ward, dictated by my impatience and my belief that everything works out in the end.

These gardens have been the testing plots for our mistakes, experiments, and crazy hopes. Sometimes things work, sometimes they don't. This garden

has been an incredible learning ground. Those first few years were a constant struggle against the march of nature. Here we had faux raised beds in a field. The grasses and subsequent white-flowered bishop's weed were not daunted by boards that rested on top of the ground, or a surface barrier of cloth and rock. In came spider grass and orchard grass. Spider grass is delightful to weed, its shallow succulent roots happy to release from the soil. Orchard and crabgrass are an entirely other entity. They mat and colonize. They arrive in a place to stay. I would make an early spring cleanup in these gardens, pleased with the turned dirt and the breathing room around plants, and within weeks the grasses would reach from the bordering field and fill the beds and crowd the residents meant to be there. I would tackle the weeding again. And again. And again. I would get waylaid by other responsibilities at the osteria at the height of our busiest season, then turn around and the gardens would be once more overgrown.

Lesson number one: Do not fight nature. You will never win. This became so apparent in that first summer of the new gardens, but it took a while for the knowledge to push in, to actually take root. At least a couple of years. During that time, we began to read and study garden making; we began to form clearer ideas of the form and function of our gardens and the way we wanted to garden. And as the gardens grew to the walled garden in which we did the work in the right order and with higher rates of success, the pressure was removed from the original potager to be everything: flower, herb, vegetable. We had more than enough space to have several different kinds of gardens, and it made sense at the beginning to divide the uses, from both a form and a function perspective.

A smart gardener knows when to admit defeat, knows when to start over. A romantic and philosophical gardener will believe that she can work with what she has and make it better. While I hope I have learned how to be smarter when it comes to dirt and green things, and I do think I have, I am also firmly in the romantic and philosophical camp. The smart move would probably have been to start over yet again with this original potager. Remove all the plants, till the beds, build up and amend the soil, then replant. But we had already done significant work in these gardens, and a number of years had already passed. The plants were huge despite the yearly battles with grasses. These gardens always had a lush and beautiful moment during their season, usually around mid- to late June when everything was in balance. It was overwhelming to think of starting all over again.

So I cut a compromise with myself. I would not dismantle the four beds, but would gradually dig them up to divide plants that had grown too big, and amend and build up the soil as I went along. I would also add two new beds, but ones that were done correctly from the beginning, resulting in a slightly terraced garden that stepped down to the winter garden between the little apple and cherry orchards. By the time this work began, I had a firm idea of what I wanted these gardens to be within the larger framework of our farm: a rose garden supported by other scented and herbal companion and kitchen plants.

Why Roses? Why Not?

Certainly, the two still-enigmatic roses already planted on our property made an impression with both their aspects and their mystery. These two unknown, unidentified roses captivated with their call to discovery. They made it seem that roses have an inherently intriguing background, and for anyone interested in plants and their evolution in history, roses appeared to be willing to give quite a ride. And so they are. They tell a story full of life, as do all plants.

WE DO NOT MAKE decisions in a philosophical emptiness, just as gardens, or vineyards, or orchards do not live and grow in a void. So many other things were happening at the time our forgotten farm became a new farm, all of which informed our thoughts and plans. When I began to re-envision what was once the potager, I became enamored of infused liqueurs. We had spent some time in the spring of the year of the garden transition in the wilds of deep southern Italy where there is a rich culture of *rosolio* and *amaro*. *Rosolio* means "sun's dew," and *amaro* simply means "bitter." While there are many recipes found throughout southern Italy for rosolio made with fragrant roses and amaro made with bitter herbs and roots, you can use any botanical worthy of maceration. I came home from those travels itching to make my own liqueurs. In my work at the osteria, I was fully immersed in my study of and work with Italian wine, but I hadn't had the realization yet that I could make or grow wine here in Vermont. It would be a couple of years before that epiphany would arrive. I think even at that time I had begun to wish myself a crafter of something liquid and surprising, something that elicited the soul of a place, for that is what the numerous rosolio and amaro tastings in Italy had done for me. I learned something unquantifiable about the people who lived there, their hopes, their concerns, the quotidian shape of their days. I learned something of their culture, which revolved around, even in this relatively industrial age, the diurnal rhythms of the land and the 365 days of the year.

I learned something from my palate and my gut rather than just my brain and my eyes. I had long been a believer in wine as an art, an art appreciated with the tongue and nose just like a painting is appreciated with the eyes and music is appreciated with the ears. Too often we forget that truth can be perceived by our other senses. Rosolio and amaro libations were just another way to express something potentially magnificent and honest.

The garden would become my palette. I would grow things to make different kinds of rosolio. Because some of the most fascinating things that I had tasted were made with roses, and I had found a recipe for rosolio in an old card shop tucked into the narrow, intriguing, and messy streets of Naples, I decided to work primarily with roses. So roses I would plant, and the hodgepodge garden that I had been growing would become a rose garden.

Varieties

Always a gardener eager for immediate results, I made several trips to several different local nurseries. I would wander for hours in their rose sections and procure as many roses that elicited fragrance as I could fit in the back of my car; I allowed what was available to me to dictate the varieties: *Rosa rugosa* in varying shades of pinks and white, and numerous bushes of Therese Bugnet, an elegant shrub rose bred in Canada to withstand the harsh winters with a fragrance reminiscent of old roses. Often the Therese Bugnet is one of the frameworks for antique rose gardens here in the Northeast. Just like in our vegetable gardens and orchard, I focused on heirloom varieties. In came a succession of Bourbons: Louise Odier, a lovely warm pink color shaded with lilac; the classic Souvenir de la Malmaison, named after the Empress Josephine's rose gardens at Malmaison and smelling strongly of tea; Madame Pierre Oger with her chalice-shaped cup. All heavily perfumed old-style roses, perfect for extracting scent and flavor in a liquid form.

ROSES AND SCENTED GERANIUMS

MY PLANTINGS OF ROSES, so dominated by the Bourbons, become heady with scent during the temperate and wet month of June when everything bursts into bloom. These foreign but decidedly English-in-character roses seem to thrive and enjoy English-like weather despite their French and Asiatic lineage. Every June, since I began to seriously plant roses, has fallen under the moniker "English Summer."

Roses in every shade from white to dark pink and even fuchsia unfurl their petals, releasing fragrance and expression. The bushes tumble, the climbers climb, others have canes that arch and trail gracefully. A succession of fat buds await the constant pinch-back of the constant gardener. On misty, slightly rainy cooler days, we might make a small fire in the woodstove in the house, and outside, those tendrils of summer wood smoke hang in the heavy air and mix with the rose's fragrance, creating something highly exotic verging on a perfume of sandalwood and myrrh. Only that sweet apple scent of the few Eglantines that line the perimeter of the gardens gives the essence of the garden equilibrium, keeping it from teetering into darker and muskier waters. This is not a shadowed or mysterious garden, but it is a garden reminiscent of *sillage*, the French term used to describe a scented trail left by someone wearing a fragrance. It comes from the word for "wake," as in the trail left in the water by a boat. This is a garden that has become something a little diffuse, yet radiant when it is at its peak. The memory lingers.

IT IS OFTEN SAID that winemakers and *maître liqueuristes* are frustrated parfumeurs, and I believe this to be true. As I planted this garden, I began to study the art of perfume as well, how master parfumeurs combined ingredients, noting which ingredients and what botanicals offered different kinds of high and low notes, how they came together symphonically in scent, how each perfume reads differently on different people because of our individual chemistries. This approach greatly affected how I would craft my concoctions, and as a consequence what I would plant in this garden.

I had the core old roses to inform my field blend, but I found I could continue to search out other old roses to create balance. The more I learned about rose varieties, the more I wanted this one and that one. Just like in grapes or apples or vegetables, the names alone were seductive. Wouldn't Frau Dagmar Hastrup be an excellent addition for her towering height and the fact that she is beloved by bees? Albéric Barbier would add a long period of late bloom with a fragrance reminiscent of green apples that would entwine nicely with the ripening fragrances of our own apples in the orchard. The Fairy would add a low, ground-trailing tumble with sweet, small cupped

A BRIEF HISTORY OF ROSES

Roses have an intricate and fertile history. Fossils found in Europe, Asia, and North America tell the story that roses have existed for about thirty million years. Among some of the earliest indications are the floral decorations on Minoan jewelry discovered on the island of Crete, dated somewhere between 2800 and 2100 BC. A thousand years later, roses began to appear in the paintings and carvings of the subsequent inhabitants of the same island. The first literary reference to the rose is in Homer's *Iliad*, when Aphrodite anoints the fallen Hector with rose oil.

The earliest cultivation of roses may have taken place in China. Confucius wrote that roses were grown in the exotic and verdant imperial gardens of the Chou dynasty. Confucius lived sometime between 551 and 479 BC. The Greeks also grew roses, but not to the extent of the Romans. In the ancient world, rose cultivation was at its peak during the Roman Empire.

Ancient Romans, extravagant by nature, were also extravagant with roses. Wealthy citizens used hundreds of thousands of rose petals to carpet and scent their floors. Nets filled with rose petals were suspended from ceilings and released during celebrations to cover guests in a gentle cascade of color and fragrance. Romans made beds out of rose petals and added the fragrant flowers to their bathwater to perfume and preserve their skin. They prescribed rose petals as medicine and used them in cooking and to flavor drinking water and wine.

Eventually the desire for roses in ancient Roman culture became so great that not even the huge shipments of roses supplied from Egypt could keep the Romans in rose petals. As a result, they began to grow their own, designing greenhouses heated by piped-in hot water to create a microclimate of warmth to grow rose blossoms even through the winter.

The fragrant oil that roses produce, known as the attar of roses, is said to have been discovered by the wife of a Kashmiri ruler in the mountain kingdoms of northern India. Legend suggests she noticed an oily film on the surface of a narrow stream that ran through her rose garden. Scooping up some of the oily and petal-laden water, she rinsed her face and inhaled the exquisite attar.

Rose oil has been used by many different cultures both as a perfume and to anoint the dead. In China,

only royals were allowed to use the sacred oils, and in medieval France, commoners were only allowed to enjoy this regal luxury on their wedding day.

By the early 1800s, Napoleon's Empress Josephine had grand rose gardens at her private residence, Malmaison, outside of Paris. Her gardens are considered the first international rose collection, and they were designed to showcase the beauty not only of the blossoms but of the plants themselves. At Josephine's death in 1814, 250 varieties of roses were said to be growing at Malmaison. Many of these have been preserved, at least pictorially, by the botanical illustrations of Pierre Redouté and Claude Antoine Thory.

Old garden roses have long held a significant place in horticultural history. Their lineage follows a path grown thick with stories of intriguing people and landscapes. The Provence Rose, *Rosa centifolia* (the rose of a hundred leaves), has been cultivated since the Middle Ages and became a symbol of Dutch and Flemish still-life paintings of the seventeenth and eighteenth centuries. The Damask Rose, *Rosa damascene*, hailing from the ancient city of Damascus, first came to Europe in about the twelfth century on the trade routes from east to west that had been broken open by the Crusades. Virgil first wrote about the Damask Rose in 50 BC.

In primitive cultures, roses were valued as an important source of food long before the cultivation of field crops. The Romans are thought to have brought the practice of flower eating to Europe. However, peasants of the time considered this an abomination. Removing the flowers prevented the formation of the fruit, which was often used in rural cooking.

In the eighteenth century, more than a third of all herbal remedies for various ailments called for the use of roses. Historically, the rose has long had a wide range of medicinal attributes, its healing properties considered to be held in the petals, particularly of *Rosa gallica*, or the Apothecary Rose. The petals must be dried immediately after picking to be effective; they were used as a tonic and for their astringency.

To strengthen the stomach and aid the digestion, a conserve would be made from rose petals or a liqueur of rose petals steeped in brandy. A syrup from roses, made specifically from the Damask Rose, used to be commonly prescribed as a purgative. Farm women here in rural Vermont used the syrup to make a gentle and tonic tea. Rose vinegar, made by adding dried petals to distilled vinegar, was believed to relieve headaches. And the fruit of the rose has long been used in medicine, the rose hip often blended with a little sugar, acting as a general cure-all.

Rose hips are one of our chief sources of vitamin C, and today they are often seen in natural food stores in teas and preserves. The most edible hips are from *Rosa rugosa* and the Dog Rose.

Most of the modern cultivated rose varieties are descended from about eight wild species, which botanists commonly refer to as *Gallicinae* or *Rosa gallica*. The wild species or natural varieties of the genus *Rosa* make up a group of extremely beautiful and intriguing garden plants. Derived from these eight sisters, cultivars like the Eglantine Rose, or Shakespeare's Rose, are prized for the sweet wild pippin scent of the leaves, and *Rosa virginiana* shows off her yearlong colors—rosé-champagne-colored flowers in spring followed by glorious dark hips in the late summer, autumn shades of red, yellow, orange, and green leaves, and the arching wine-red canes of deep winter.

Even though all roses trace back to a fairly small family of origin, the evolution of the rose is quite intricate and convoluted. When the China Rose was brought to Europe at the end of the 1700s, a sort of rose revolution took place. Until then, the only roses found in Europe were hardy shrubs that only bloomed in late spring and early summer. The China Rose radicalized the European rose, bringing with it not only novel colors, like yellow, but also the ability to repeat bloom, trail, and climb.

The European China Rose as we know it has been traced back to four stud roses that arrived on Western shores. The first was Slater's Crimson China, imported by a gentleman horticulturist named Gilbert Slater of Knot's Green, Leytonstone, England, in 1792. By 1798, the French, who dominated rose breeding at that time, got their hands on specimens and began hybridization experiments. Within a few years, rose plants derived from the Slater's Crimson were found in Austria, Germany, and Italy. And only now is it supposed that a very closely related cultivar has actually been in existence in Italy since the mid-seventeenth century.

The pink China Rose is thought to have come to Europe even earlier than the Slater's Crimson, via Sweden in 1752. The first mention of this rose occurs in the nursery paperwork of Englishman William Malcolm, who noted it as Evergreen Chine and a "new Chine." Years later, he added a botanical name: *Rosa indica*, the species name actually referring to China. The first mention of this rose actually as Parson's Pink China in England occurs in 1793 when it was being grown at Rickmansworth in the gardens of Mr. Parson. Such was its impact on gardening in England that by 1823 it was said to be in every cottage garden in the country.

Parson's Pink China appears to have been officially introduced into English horticulture in 1793 by Sir Joseph Banks, then director of the Royal Botanic Kew Gardens in London, and was most likely collected near Canton by Sir George Staunton, a member of Lord Macartney's embassy in China in 1792. James Colville propagated and sold it under the name of Pale China Rose and later it acquired the name Old Blush.

The other two China studs were Hume's Blush Tea-Scented China introduced by Sir A. Hume from the East Indies in 1810, and Park's Yellow Tea-Scented China; both most likely were brought to England by John Reeves, who was chief inspector of the East India Company at Canton from 1812 to 1831 and was responsible for the introduction of many Chinese collector's plants in Europe at that time.

With the Asian line, an increasing number of hybrids and crosses began to appear. Through the nineteenth century, a very active time for horticulture in general, it was common practice to plant naturally pollinated seeds, and for this reason the predecessors of now common cultivars can only be traced through the maternal side. The hips of many cultivars were gathered at random and, as a result, many have lineages left to conjecture.

In my own garden, Bourbon Roses began to reign supreme; they are one of those cultivars that has a murky history. Bourbon Roses supposedly arose out of natural and chance crossings between several different types used as hedges to separate land plots on a very small island in the Indian Ocean.

The island, initially discovered by Arab sailors as early as the third century BC, was volcanic in origin, with fairly constant eruptions, which probably discouraged settlement. The Chola Navy of one of the longest-running ancient dynasties of India arrived there in the eleventh century and christened it with its first name, Theemai Theemu, which means the island of destruction. The Portuguese arrived several hundred years later in 1635 and found it uninhabited, but they renamed it St. Apollonia after the virgin martyr who had all her teeth knocked out by an Alexandrian mob in the mid-200s BC. By 1649, the island was taken by the French and named Île Bourbon after the French royal house. In 1655, colonization began when the French East India Company sent twenty settlers, and the island was renamed once more: Reunion. It is still a French island with over eight hundred thousand inhabitants, located east of Madagascar and west of Mauritius. A tropical island, the spring and summer are temperate and dry, while the fall and winter are hot and wet. Reunion currently holds the record for highest rainfall in twenty-four hours. On January 7, 1966, the island received 73.62 inches of precipitation.

Bourbon Roses probably began as a natural cross between the Parson's Pink China and the red Tous-les-mois, or even possibly Quatre Saison, both Damask Roses. And all were used as hedge material to separate properties on the island back in the seventeenth, eighteenth, and nineteenth centuries.

A. M. Périchon, a French botanist living on the island in 1819, found a specimen from seedling roses raised for a hedge that was quite different. He singled it out and planted it alone to watch how it evolved. When it bloomed, it was clearly not like any other rose on the island. He showed this new rose to the head of the island botanical gardens, M. Bréon, who was enchanted by this mystery rose and through examination deemed it to be a natural cross between the Pink or Old Blush China and the Damask. M. Bréon sent seeds and cuttings from this new rose to the head gardener for the Duke of Orleans at Neuilly, the renowned rose breeder Antoine Jacques, and from there the first official Bourbon Rose blossomed in 1821. From this single unique plant sprang a whole catalog of Bourbon Roses.

Almost two hundred years later, I have a collection of Bourbon Roses in my own garden in the small state of Vermont, a landmass once inhabited by volcanoes from the Mesozoic Age, a current climate that is definitely not tropical but is characterized by hot, humid summers, and very cold, alpine winters. The highest recorded twenty-four-hour rainfall here was the 11.23 inches that fell in Mendon during Hurricane Irene, which hit in the year of 2011.

CHIVES AND IRISES IN THE ROSE GARDEN

ROSES, ALLIUM,
AND LADY'S MANTLE

clusters late into the autumn as well, though they would sadly offer no scent. Belle Amour, discovered at a convent in Elbeouf, Germany, would add spice to the nose with a native bitterness said to resemble myrrh. And how could I turn down Boule de Neige with its white outer petals curling in, its dark green glossy leaves, and a scent like sugar on snow?

Planted with the roses are traditional companions: scented geraniums and *Nepeta cataria*, commonly called catmint. There still remain the beguiling Hyperions and native orange lilies, and a feathering of white goose-necked loosestrife. A one-time gift of phlox in white, magenta, and purple has remained as the large daubs of color in the garden, offering something later in the season when the roses have become mute. There is still a smattering of odds and ends that crop up here and there: the stanchion of bright yellow golden glow, tenacious rudbeckia, and a blue bellflower, one of the first things I ever bought in my early attempts at the garden. There are clusters of peonies in magenta and blush, inheritances also, as they make lush companions to their sister roses. Clouds of chartreuse lady's mantle bloom and hang between the roses in July, offering a startling contrast. Blue veronica is everywhere, both in and outside the boxes. I once thought her to be a perfect ground cover for a naturalistic garden, until she decided to get too greedy. Every year we dig her out, thick mats of earth and blowsy leaves and stems, but she always returns. She did do her initial job well: She ran the grasses right out of town; she has now replaced their unwanted presence with her happy squatting. She is beautiful in small, delicate blue blossoms in May, making this garden at the center of the orchard look like an impressionistic painting. Later in the summer, when the rose garden overtakes itself, it will belong more to the wild brushwork of Les Fauves.

Somewhere along the line, a horticulturist friend brought a gift. A sweet little rose with double blossoms on softly arcing canes, the leaves a soft moss green. The palest of pinks, like the pearly inside of a conch shell, with a compelling perfume. A gift from his mother's garden, a one of a kind named Belladonna. He knew I was developing this garden in order to eventually harvest petals and hips for liqueur, and he thought the demoiselle would be a perfect star in the constellation of scents and flavors. He also suggested the idea of making single-variety rosolii showcasing the individual expressions of each rose.

The rose garden is still maturing, and I have continued to make a field blend in both liqueur and a syrup that we use at the osteria and when we cook for guests here at the farm, but at the back of my mind remains the notion of the single varieties and how lovely it would be to have each rose liqueur lined up one next to the other in clear glass apothecary bottles, the distillation of unique attar, a pure essence of scent and taste.

ROSE SYRUP

To make a simple rose syrup for tea, to drizzle on ice cream, or to add into a savory recipe like duck confit with rose syrup, boil 1 part water with 1 part good organic evaporated cane juice with a healthy dose of rose petals. As an example, 2 cups of water boiled with 2 cups of the evaporated cane juice and 2 overflowing handfuls of rose petals would be a goodly amount. The more rose petals you use, the stronger the flavor and aroma.

To prepare the rose petals, pick your most fragrant roses at midday when the sun is highest and the flowers are releasing their essential oils. Pick the petals from the hip. Trim away any white from the edges of the petal that was attached to the hip, as this part will be bitter.

Once the sugar and water have completely coalesced in the heat, the sugar dissolved and become liquid, take the pot off the heat and let the mixture cool. Strain the petals from the syrup. Pack the syrup up in a jar with a good lid and refrigerate until ready to use.

Lesson Learning

While the rose garden is indeed beautiful when it is in full bloom, its trajectory is not necessarily one to copy, even though I think its history makes its fruition all the sweeter. A rose garden, or any kind of flower garden, whether planted for edibles, scent, or ornamentals, still needs all the same care as you would put into a vegetable garden. Perhaps I am so attracted to roses because they are quite similar to grapes in their needs and wants. There is a long history of planting roses at the heads of vine rows, as roses are susceptible to many of the same ailments as grape vines, but they will get these maladies a little earlier, giving the winegrower notice that complications may be on their way.

BEFORE PLANTING ROSES, work the soil, amend with compost, aerate, then plant. Roses will grow well in many different kinds of soil, but loam is the perfect habitat. Rose plants need water, but also air and space in the dirt around their roots. Sandy soil drains too well, and clay soil retains too much water and can be confining. Something between the two will make the roses happiest: a balanced loam. Roses also like a slightly acidic soil leaning toward neutral.

Just like in any vegetable garden, it helps to think ahead. To break ground the year before planting on a brand-new bed is wise. I wish I had given myself at least a couple of months to work the soil and add in compost before planting; this allows the new organic matter to settle in and marry with the original soil. If the soil does not drain well at all, if you are clay-heavy, a raised bed will be the best choice.

Roses will respond well when planted in prepared beds with either well-rotted manure or garden compost, and once they are planted, they will need a steady diet of nutrients during the season, especially if they are repeat bloomers. Roses love vegetal compost, or a slurry made from nettles. This is the perfect compost companion to any kind of rose.

Roses also do well with organic mulching. Straw or wood chips can help conserve water, keep the soil cool, and feed the microbial activity.

Pruning

P runing for roses is not unlike pruning for grapes or orchard trees. Late winter and early spring are the best months to approach your plants. Pruning work should ideally happen in March and April, when the weather starts to warm slightly but the plants are still dormant. It is best to prune well before bud swell but after heavy frosts. This timing can be complicated for New England roses, as frosts can happen as late as the end of May, but if cold-hardy varietals are selected, this will go a long way to protecting them. Pruning helps the plants keep their shape and keeps them flowering, and just like with grape vines, it's important to open the plants up to light and air, especially when following an organic paradigm in our part of the world.

PRUNING IS CRUCIAL in keeping the plants shaped and blooming and fitting into your garden space, but it does not have to be an exact science with roses. Grape vines are the most particular because you are pruning for a certain kind of crop; orchard trees can be too, but they seem more forgiving of mistakes. And roses are the most forgiving of all. They will bloom in some fashion whatever the circumstances. Pruning just helps them keep healthy and ordered.

When pruning a shrub rose, take it down to about half its size, giving it a simple, rounded shape. Clear out any dead or weak-looking wood. The only time to vary from this pattern is maybe in the first year when the bush is still establishing itself. In that case, give a lighter prune, keeping about two-thirds of the original. Then, from there on out, prune back half. The size of the bushes can be managed in this way. If the goal is a smaller plant, there will be no harm in pruning down to one-third of the size, and if the want is for something larger, address the rose as if it was in its first year. A typical English planting would place three rosebushes together in close proximity for a mass effect. All three can be pruned as if they were one shape.

We have climbers and ramblers both in the garden and around the farmhouse. Near the house they climb straight up pergola supports as well as frameworks on the wall. When planting a rose to climb up a single pole or arch, make sure it follows in the right direction. Tie the rose to the support to encourage it upward. Until the rose reaches the height wanted, prune minimally. Just trim out any damaged growth.

As the plant reaches for the designated height, and once it arrives there, new flowering stems will start to grow out from older wood lower down in the plant. During the summer season, it's a good idea to cut these new stems back to three or four sets of leaves after flowering to encourage continued flowering; then, at the end of each season, cut out any dead or weak stems and tie newer stronger ones. Otherwise, the plant will eventually get straggly and exhausted and the production of new growth and blooming will slow down.

The effect of a rose climbing up a wall calls for a slightly different method to get the new growth to eventually fan out and cover the wall or structure. Instead of just tying the plant upward, tie some stems outward as well. The closer the stems are to horizontal, the higher the number of buds that will develop into new flowering stems. The idea is similar to creating the espalier of a fruit tree. Once the basic framework is achieved, and after the new flowering stems have bloomed, summer-prune them back to about three or four sets of leaves. In the winter, the rambler can be reshaped, removing very old, damaged, or weak stems and replacing them with new growth coming from the base of the rose.

Unlike some other plants, roses respond well to summer-pruning. Deadheading is an important task if you want to keep your rose blooming. Nip off the spent hip, or cut the stem down to the first leaf. This is particularly good to do in milder climates where roses are not quite as vigorous; they tend to expire a bit in the heat.

Many roses will grow taller and taller every time they flower, and they are summer-pruned for the next round of flowering. To keep the rose from getting more height, be sure to summer-prune rather than deadhead at the end of each full flowering. This will continue to encourage the repeat blossoming, but will also keep the bush in check.

Sometimes a very strong, new long stem or cane will grow from the base of the rose and outstrip the other stems and make the plant look a little off kilter. This is actually a very beneficial stem and should be kept for good flowering the following season. This stem, or cane, is similar to the bull cane on the grape vine. While summer-pruning, trim this new leader slightly, just to keep it within the shape of the bush.

Our farmhouse and the rose garden are parallel to each other within the orchard. They form two centers, one a living house, the other a vibrant garden. The way our cultivated landscape has evolved is based on this form and structure. The farmhouse is encircled by the arms of the walled garden; the rose garden is surrounded by apples and sour cherries. At the top of the rose garden is the lamb house, and below it is the structure for the winter garden. There are gravel paths that intersect and guide in each private world, and well-trod paths through grass studded with white clover connect these seemingly separate environments into one orbit. Each garden is a satellite of the other, influencing and sharing.

But the shapes of these separate gardens are not rigid. One flows into the other, flora and fauna wending their way from one to the other. In the vineyard the rose garden extends its reach with plantings at the head of many rows. Fruit trees surround the roses, but they also find themselves in the center of the raised beds in the vegetable gardens. Young espaliered apples begin to form a wall on one side of the rose garden. We've begun to grow slightly less hardy grape vines in the winter garden, planted at the base of each metal rib to entwine up the pole and provide fruit for wine and shade for the summer plants. Grape vines twine up the terrace supports onto the balcony and anchor the stone wall, offering fruit for dessert or sweet dark jelly made after harvest. Greens, root vegetables, squashes, and melons find their way into the vineyard, and greens make themselves fine companions at the base of fruit trees. Onions and chives march into the rose garden, plants

from the onion family an ancient companion still paired together in the rose gardens of Bulgaria. Allium plants deter pests like aphids because they confuse them with their scent. They also help roses combat fungal diseases like black spot. Herbs also make good companions for roses: parsley, thyme, and scented geranium. They fill the beds with their bright fresh green color and perfume.

The gardens weave in and out of each other bringing birdsong, ladybugs, and bees. The blue-black crows in the rookery on the hill behind us bring their young to our meadow to teach them morning lessons. They swoop between trellis posts in the vineyard and up into this or that stand of birch, instructing and cawing all the while. Hummingbirds share rose nectar with

the bees. And there are so many different kinds of bees: wild bees, honey-bees, Italian bees, bumblebees. Their varied but similar suits of black and gold stripes make them sartorial in flight.

Our farm is at its most magical seen in this light, and it is magical in its own way despite, or maybe because of, its untended areas and flawed projects. These various gardens and yards that nestle in against one another are in no way perfect. Some years the roses have black rot, or the Japanese beetles try to decimate everything. Grape flea beetles arrived in the vineyards this year, and beans didn't germinate until the fourth seeding. These are the corrections of nature, informing us of what needs to happen next under our care. And while a pathogen or an insect may rule for the day, the apples will be perfect in their red blush, or the escarole will be abundant with near-perfect green leaves. The daikon will be harvested without a blemish and the tomatoes will ripen well into October. In the polyculture of a diverse farm landscape, all is never completely lost. For every minor tragedy there is always a victory.

I often think of my lady benefactor and her garden that has morphed and moved and changed and that is now mine, that legacy of daffodil, rose, and wildflowers that she left behind, or that are now gone. The asparagus bed finally petered out, and we have not replanted one. Yet. The few remnants

that were left, which have been transplanted time and again, some thriving, some fading, are a constant communication with her and this place we call home. Those two roses that have now proliferated and offer their brief blossoming at the beginning of the summer, that inspired a whole larger and constantly evolving garden, offer a thread to the history of our landscape—who has lived here and what they did. We are now a part of that history, continuing to shape this geography in some way still connected to our predecessors. The little (and big) reminders of her garden—the roses, the Persian coneflower, the recurring daffodils, a lone white foxglove—are like a sillage of the gardener. As are the inherited plants from our two families: the yucca from one grandmother, the lilies from another grandfather, the yellow tansy or the sweet clump of blue Jacob's ladder from our mothers. The wake of our ancestry, expressed in plants, and these pieces of our families go on through us in new and fertile ground.

Apiary

"LET THERE BE GARDENS TO AMUSE THEM,
WITH THE SCENT OF BRIGHTLY COLORED FLOWERS . . ."

—VIRGIL ON BEEKEEPING, *Georgics*

I believe the perfectly functioning farm has animals. There might be ducks or chickens for eggs and meat, a cow for milk, a horse for plowing, sheep or goats for milk and meat as well, bees to make honey, cats to catch mice and prowl the gardens and greenhouse, dogs to watch the sheep and goats, and the dogs, again, to watch after their masters.

AND IN A PERFECTLY FUNCITONING FARM each part of the puzzle feeds the next part. The manure from the cows, sheep, goats, chickens, and ducks feeds the vegetables, grains, flowers, orchard, and vines. The vegetables and grains feed the chickens and ducks; the hay feeds the cows; the grass, insects, and wildflowers feed everyone; and everything feeds the farmer. And the dog.

Subsistence farms have always had some kind of livestock, even if only a few chickens for eggs and the occasional roast. It is no secret, whether you practice biodynamic or organic agriculture, or even farm more conventionally, that the animals bring all kinds of certain energy to the land. Their manure brings energy to the plants and soil, replenishing what is absorbed each year. Their meat brings energy to us as carnivorous humans. But they also bring the animalistic energy of their daily lives.

Modern small farms are sometimes faced with choices. Many small farmers must take other jobs to keep the concern going, and their schedules do not support the care of larger livestock that must be let in or out, barns that must be cleaned daily, animals that must be milked. Modern farmers often have to have work lives as diverse as their farms in order for their farms to be economically sustainable. Smaller livestock like chickens tend to be easier, and they can fend pretty much for themselves throughout the day if they have run of the farm. Our neighbors would let their chickens out in the early morning when they would collect the eggs, then call them into the coop again at night before going to bed. This was in the summer. Winter hours are more complicated, as the birds can't go out so easily when there is deep snow. They must be fed and watered regularly. And, where we live, there is the added threat of wild predators at the ready on the edge of the meadow. Fox, coyote, mink, fisher cat: They are all glad to partake of a chicken dinner. They are greedy; rarely do they leave a coop with survivors. At the neighbors, this has happened twice. They now no longer have chickens.

So there are questions of fences and protection (those dogs) and heartache if indeed Mr. Wile E. Coyote should enter the hen pen under your watch.

One hundred years ago, those questions weren't asked; they were simply part of life on the farm. Farmers were at home, they were growing their own food, and they would sometimes lose animals, just like they would lose crops, the often violent mood swings and routines of nature just as relevant in this part of the cycle. But this is not one hundred years ago, this is today and there are new questions and the new concerns.

We, ourselves, do not have livestock on the farm yet; a large number of vineyards and orchards do not. The farmers' energy must go into the tending of the vines and trees, leaving little room for the care of animals. Because our farm feeds our osteria (that second job modern farmers might take), our daily schedule currently only has space for the plants and two cats.

But this question of the perfectly functioning farm has nagged at us, the belief that animals form a more perfect, holistic circle, and ultimately make a farm complete. We grapple with what kind of livestock we could properly care for and raise. Animals, while so important to the farm, represent big change. Change is the end result of transition from no animals on our farm to animals living here in tandem with us. We've been looking for a gateway. It occured to us that opening the animal door to our farm could be heralded by honeybees.

We set out to learn more about bees. In the thick of January on a cold, snowy, sunny morning, our friend Michael took us to see his bees. I've come to think of him as the poet-beekeeper, for he has been a poet and teacher for many years, and now he is also a beekeeper.

We stayed overnight, traveling over hill and dale to get there for a dinner of piquant and seductive flavors of a New Year's New Mexican feast, a tradition born out of his wife Katherine's childhood. The night was cold with gale-like winds buffeting the house, but we were snug as if in a small boat that somehow stayed centered on the sea.

The morning brought Persian eggs, a scramble of egg, peppers, onions, and spices, a recipe from another poet friend hailing from the Middle East, and who is no longer, the making of this dish a remembrance. And the morning also brought talk of bees. When Michael had made a gift to us of his first honey a couple of years before, we became enamored of the idea of having our own bees, and in the time frame between that first translucent jar of honey and this winter visit, we had finally decided to bring bees to the farm. That notion that we were without livestock, always dogging us, had brought us to the thought that bees could be the perfect solution, and bringing their energy and productivity to our small vineyard, orchard, and vegetable gardens would be one way to bring the animals' life to our small farm.

Ever since my experiences as a child, I've loved the idea of bees, the way they looked, moved, sounded. Since living here on Mount Hunger, I have loved the bees that come down the hill from the old farmer's son who kept hives up above us, loved watching them take nectar and pollen from our abundantly blossoming fruit trees, their black-and-yellow-striped bodies twitching with the delight of their tasks. Bees also played a role in the early part of Caleb's and my connection to each other, at least as the fantastical creation of a country wedding: Our celebratory cake had been fashioned into a beehive, an extravagant dome of lemon cake covered in a lemon whipped cream with a portal of cascading woodland berries and dotted with honey-bees made from marzipan and almond wings and festooned with brightly

MEMORIES OF BEES

My first memory of bees is exceedingly clear. I am four years old. It is summer. Late July when the Queen Anne's lace has started to bloom and the air is full of the flower's heady, honeylike perfume. We are in Vermont. My parents have rented a house for the summer, here in the same village where I would later come to live and open a restaurant, start a farm. But who knew any of that then?

My father is a teacher and educator; while I was growing up he had summers off. My mother had gone to school in Vermont, and my father not too far from here, so they decided one year that it would be good to get out of the debilitating southern Indiana heat and head north to a place they both remembered well. We packed up two cars: a white station wagon that had seen better days with an equally worn horse trailer attached to it, and a snappy dark blue Volkswagen Beetle with a gray convertible top. There were: two parents, three daughters, two dogs, six cats, two ponies, and a parakeet named Danny Boy, though not because he could sing nostalgic Irish melodies. He was more given to expressions using colorful four-letter exclamations.

We took three days to travel from the shores of the Ohio River, two overnights. At the overnight motel stops, the cats had to have their own room. We would hide the dogs while going in and out of the motel, usually with my father's raincoat. Difficult work with the likes of an enthusiastic Irish Setter.

We arrived to a beautiful Vermont summer with hot, sunny days and warm rains. The house where we stayed was old and clapboard, but done up nicely in fresh white paint. The rooms inside were colored variations of haint blue—that porch ceiling paint—giving the indoors a languid, sleepy feel like a summer porch. In the Gullah culture of the Low Country in the South, they believe that spirits, or "haints," can't cross water. They used to paint the porch ceilings and house doors a pale blue color thinking it would deter any nearby ghosts. Luckily, it also is the color reminiscent of the sky and seems to confuse mosquitoes. In this Vermont summer house with its soft, muted rooms, the beds were all dressed in white, nubby, chenille bedspreads.

The house had a particularly striking feature: a grand staircase built on the outside of the structure rather than the inside. Two sets of elegantly curving stairs stepped down from a second-floor balcony, their shape evoking two arms embracing the house, the lawn, and you. The front door was situated under the center of the balcony, beckoning you to come inside the cool, watery interior.

My sisters and I used to run up and down the stairs from the upstairs bedrooms that looked out from the balcony, the novelty never wearing thin. One of the arms, on the right when you were looking straight on at the house, held a huge wild beehive. All wattle and daub, it looked like a storybook hive, perfectly round at the top like a hot-air balloon, and pointed at the bottom, funneling the bees out one by one when they left to go out and collect pollen.

We usually used the other staircase, so as not to disturb the hive. There were strict orders about that. Especially from the local gardener who tended the property while we rented and the home family was away. One early evening, my father was taking photographs with his camera, a little Brownie that documented all our family travels. Artistically inclined, he thought a photograph with one of his daughter's next to the near-perfect wild hive would

be particularly fetching, especially since the hive was bigger than she. I was the youngest.

My mother was not nearly as keen about this photo shoot, admonishing my father to take a photo elsewhere. But he said there was no need to worry about the bees. They had been docile all summer, and it wasn't as if we would be bothering them. But my father had never been a beekeeper; his gentlemanly farming experience veered more toward livestock and vegetable gardens.

So up the grand staircase I went to pose discreetly next to the hive. I remember the hum of the hive sounding warm and intense. Maybe I began to be afraid, which raised the call to the hive. I remember the bees starting to fly out from the opening of the hive. I remember thinking it was time for me to come down the stairs. I remember my father saying, "Just one more photo!" as he was too immersed in looking through the lens to see the accumulating swarm of the bees.

While I was afraid because I didn't understand what was happening, in reality, the swarm was more comforting than terrifying, and it would have probably been fine if I hadn't opened my mouth and yelled my confusion or tried to move. But I did, and there I received the bene-diction of the bees, a series of stalac-titic bumps on the roof of my mouth where they stung. There were no other stings.

I ran, down the stairs. The bees quickly left me alone. There was ice for the swelling, and salt from the tears. Ever since that time, I have felt the mark of the bee, a fascination with their culture, their physiques, their sound, their honey.

colored flowers from my father's dahlia garden. Bees weave their lives around us all without us necessarily knowing it; without them our food chain would collapse. But in my life and my life with Caleb, bees seemed to follow us wherever we go.

Michael had fallen under the bees' spell a few years before us, and had started to bring us that honey that so inspired us every year, a delectable mélange of complex flavors from the wildflowers growing in the hills of the Champlain Valley around him. One year he brought a dark, smoky, melan-choly honey from a hive that had died, the taste so poignant and haunting, the final work of these bees gracing our mouths.

On that bitter January morning at Michael's side, we leaned down close to the hives to listen at the wall. Knees in snow, ears pressed against the painted wood, there was a low thrum that vibrated down through the skin of my earlobe, down through the aural pathway, and into my chest. A subtle yet thrilling thrum that woke me up to the metallic scent of the air, the feel of the weak sun on my face, the pinpoints of cold in my feet, knees, and hands.

nu·cle·us /ˈn(y)o͞okleəs/

1. *The central and most important part of an object, movement, or group, forming the basis for its activity and growth. (n.)*
2. *The positively charged central core of an atom, containing most of its mass. (n.)*

As I look at the hive nestled against a frothy pine tree at the top of our vineyard, I will preface that our relationship with bees is still at the very early stages. Compared with other aspects of our farm, we really sometimes feel the want of what we don't know in the world of beekeeping. We are students in all areas of our farm; I believe that all good farmers are always and forever students. But like so many things we've done, we dove into beekeeping without doing a lot of study and preparation beforehand. Certainly, we collected some information, but as is always in the wisdom of hindsight, I wish we had taken more time to read those manuals stacked by our bedside, the old beekeeping books from Caleb's grandfather's farm in New Hampshire, and I would advise anyone who wants to start working with bees to do their homework first. Not that bees are difficult, really quite the opposite, but there are many nuances and aspects to ponder before the bees arrive, and to be sure you give them the best care in their new home, you must put some forethought into your upcoming life with bees. To twist another saying, preparation is the better part of valor.

How to Bring Bees Home

In the spring after our winter visit with our friends Michael and Katherine, we arranged for bees to arrive at the farm. There are several ways to acquire bees (though this suggests that you own them, and, wild and instinctive by nature as bees are, no one ever *owns* their bees). Michael had suggested working with someone local who raised bees. He had the advantage of a neighbor not far from him who raises bees and was very experienced and willing to mentor Michael in any way he could. We wanted the same kind of relationship. We knew our neighbor at the bottom of hill had just started his first two hives the year before. We went to him first.

A HONEY BEE NUC

DEAN ALWAYS MODESTLY says he is not well versed with bees, but he was willing to help us get set up and connected us to another good local person. Dean may not have had a lot of years working with bees, but he has a naturally calming way about him, and his uncle was renowned in our village as something of a bee whisperer: a man who could lead a wild, flying swarm within the semicircle of his own arms to a new manmade hive. Those who saw his embrace of the fluid and animated shape the colony of bees made call it nothing short of miraculous, a moment where a man and the wilds of nature are one.

The bees arrived at the farm on a warm Sunday morning; Dean brought us a nucleus colony and a complete hive setup to get us started. He and Caleb put the nuc (something that looked like a wooden shoebox with a hole in the side covered with a very fine mesh that could be opened and closed to let the bees in and out until we got them situated in their new hive) in the north orchard, away from activity but in sun and under the branches of the apple trees.

Getting a nucleus colony was undoubtedly one of the best and easiest ways for us to come by a hive, and it is always recommended in manuals and by beekeepers. You can also receive bees in a package through the mail, but many manuals ultimately discourage this if it's at all possible to find the bees locally. In the postal package, the bees will be cooped up in quite small circumstances, and you will have to be ready to get them settled immediately as it only begs for trouble for them and you to keep them so tightly confined. In this scenario, a queen will also have to be bought separately, whereas when a nuc arrives the queen is included. Buying the workers and the queen individually is more expensive, and for the novice beekeeper, getting all this organized is more difficult. However, it can be done, and if a decision needs to be made between getting package bees or no bees at all, mail-order is better than nothing.

Bees raised locally have already adapted to your particular landscape. Bees that come from farther away will take longer to adjust. We started out with a nuc, but you can also buy a complete, fully functioning hive from a beekeeper who is downsizing or retiring. There are many positive advantages to buying either the complete hive or the nuc. With the hive you have bees ready to go. You don't have to worry about moving frames or introducing a queen or getting a hive ready. You just find the proper location. But there are some potential downfalls. An established, older hive can have diseases, pathogens, antibiotics, and chemicals present in the wax of the comb, even when the beekeeper is adamant about not having used chemicals in his beekeeping. Bees collect more than pollen when they are out in the world; all kinds of materials in the environment can attach to the bees, which they then bring back to the hive and the comb.

When buying an established hive, it's important to look for the signs of things like varroa mites, hive beetles, chemical residues, and disease pathogens. If the comb is dark and not marked, this is a good indication that your beekeeper is unloading an older hive that will inevitably carry some of these problems. A lighter-colored comb suggests a newer hive that has yet to weather the vagaries of bee life. Take a friend or mentor who is knowledgeable to help you check for the telltale signs and adequately judge the state of the hive.

Getting a complete hive is also a heavy lift—the weight is simply cumbersome to move around. Buying a nuc is easier: They are smaller and lighter and easier to manage, with many of the same advantages as the complete hive. You get frames with comb and a full assortment of the bee family, including the queen. You slide the frames into the new hive, and the bees are essentially moved in and ready to go. When bees come in a nuc, you can also get a better sense of how the bees were raised and how "new" they are. If they are coming on pre-established frames, you can see the color of their wax just like in the complete hive; the darker the comb, the more certain you are getting an older and possibly contaminated frame.

It may not be possible to avoid getting a frame with other elements attached, and if this is so, it's important to rotate them out the following spring as soon as possible and give the hive a new frame with a new foundation.

For the beginning beekeeper, understanding the care that the previous beekeeper has given to the hive or nuc colony is extremely important. Have they used chemicals? Are they organic? How have they split up their hives when it's time to do so? Are the bees at risk in any way? If the nuc comes from a healthy hive, the clearest concern is how to continue keeping them healthy and happy.

Varieties, Species

To look at an organic beekeeping regimen like an organic farmer looks at his or her land is to focus our gaze on the hive. There are many divergent views in beekeeping, some supported by industrial agricultural paradigms, in which our human interests in producing the greatest amount of honey our hive can muster are paramount; in this system, the type of bees and how we micromanage the hive allow us to extract the most bounty. Other views center on a more natural husbanding belief, that we all have a role to play in our larger cosmos. Bees are just as important as soil, trees, or humans. All of us live together in an intricate web of life that operates by an environmental design, and our goals in keeping bees should be more along the lines of what is beneficial for the bees rather than what is strictly beneficial for us regardless of the bees' needs.

I BELIEVE OUR ROLE AS BEEKEEPERS is not to extract the most honey based on our needs and desires, but rather to care for the bees in a way that allows them to live healthy and productive lives that feed them, first, and give us some extra honey as a by-product. Working with bees and hosting bees on your land does so many other things besides encouraging honey production. The bees provide necessary functions in our chain of life. Their act of pollination feeds us and the other animals that populate our world, many of whom also in turn feed us. Without bees, the basic foundations of our existence would eventually crumble.

Choosing the species of bees that come in a nuc or package or hive is a lot like choosing the varieties that will be planted in a vineyard or garden. You must take into consideration the terroir, and specifically the climate, landscape, and your own level of experience.

There are several different varieties of bees available to us here in the United States: Italian bees, Russian bees, Africanized bees. Some are more docile than others. I'm sure you've heard terrifying stories of killer bees, or seen the horror movie. Africanized ("killer") bees are a type of bee that is particularly aggressive, but if handled properly and by an experienced beekeeper, there is nothing terrifying about them. Russian bees are good for cold-weather climates, but are still a bit aggressive. Italian bees are the most docile and the easiest to work with for the beginning beekeeper. Our first nuc came with Italian bees.

Italian bees, *Apis mellifera ligustica*, are a species originally found in Italy. They are thought to be native to the part of Italy that lies south of the Alps and north of Sicily, the region of Liguria, and, as a subspecies, there is good indication that they survived the last Ice Age occurring in Italy. They are genetically different from the subspecies of bees found on the Iberian peninsula and those in Sicily. They are the most widely raised honeybees due to their intrinsically gentle nature and their ability to produce a large amount of honey. They have proven adaptable to most climates, from subtropical to cool temperate, but they thrive less readily in humid, tropical regions. Given that they are from the warmer climate of the central Mediterranean, they are noted as doing less well in "hard" winters and wet, cool springs. They do not form as tight a winter cluster, and more food has to be consumed to compensate for the heat loss in a loose cluster. However, because they are so mild in temperament, they are often suggested as the bee of choice for a new beekeeper. They are widely available throughout Vermont and New England, and they do survive our colder climate with a bit of preparation and assistance.

The Italian or Ligurian bee was first introduced to Britain in 1859 by the respected beekeeper and journalist Thomas White Woodbury, who lived in Exeter, Devon.

In 1859 Woodbury imported a yellow Ligurian queen from Mt.

Hermann in Switzerland. She arrived by train on 3 August in

a rough deal box with about a thousand worker bees. Woodbury
had prepared an 8-bar hive, including four frames of honey and
pollen plus one empty comb, and he gently shook the newcomers
into this. Then he took a skep of local black bees weighing 34.5
pounds and shook them out in clusters on four cloths spread out
on the grass; helped by his friend Mr. Fox. He found and took out
the queen, before placing the hive with the Ligurian queen and bees
over the shaken bees. Alas, they fought, and in the morning there
were many dead bees, but he hoped for the best. By 17 August, great
loads of pollen were going in, and he knew that the first queen from
outside Britain had been introduced.

RON BROWN, *Great Masters of Beekeeping*

As in any endeavor, the importance of starting off with the tools and circumstances in beekeeping that fit your level of expertise is paramount. Beginning with Italian bees makes the first time you do everything with your hive that much easier. Just like any other animal, bees can sense fear. If the apiary has a race of bees that is particularly belligerent, the beginning beekeeper will undoubtedly always be tentative and a little afraid of being too afraid around the bees. Fear and a lack of confidence will transmit to the bees and may incite them to be more antagonistic. It is wiser to start with more docile bees.

However, just as in any other aspect of farming, diversity is important. A diversity in the bee yard is very much desired in organic beekeeping. Raising bees of mixed genetic strains that have pest and disease resistance is preferable and can give the bees a leg up in confronting challenges.

Many bee breeders are intent on keeping a purity of strain in order to maintain a specific breed, especially queen breeders, but the beekeeper who keeps pure strains in all of her hives in the bee yard is practicing a kind of monoculture, and just like single-crop farming, this creates a vulnerability to disease and climate fluctuations, as well as a potential lack of genetic fortitude. These are important considerations for apiculture in our part of the world. Some breeds do better in colder climates than others, and most northern apiculturists notice that package bees from southern states and California, as well as overseas, tend to do less well on our ground. Even though they

MOVING THE NUC

may be touted as being resistant to diseases and mites in those locations, it doesn't mean they are resistant to the diseases and mites that are on offer here. Bees in the warmer climates work all year round, whereas bees in the North must hunker down for the coldest of winter months. Bees that need to survive the struggles of the cold need to be made of sterner stuff. Africanized and European bees tend to know how to handle the colder temperatures. The hive gives birth to an autumn brood that will live in the hive until the spring, unlike their summer counterparts who only live for five or six weeks, becoming exhausted by their long list of hive chores. The autumn bees have a different function: the warming of the hive and the queen. The bees actually cuddle up to one another to form a dense ball around the queen and to keep one another warm. When the bees on the outside of the huddle get too cold, they slowly move toward the queen in the center to get warmed up, the warmer inner bees moving to the outer layer in response. In this way they regulate the heat within their formation. They don't heat the empty space around them, but they keep the heat going within their active but quiet winter cluster.

There are many other amazing things that these winter-hardy bees do to survive during their dormancy, the brilliance of their patterns and instincts showing us the myriad and miraculous in nature. And because northern beekeepers need to promote bees that can weather our northern cycles, raising successive generations of bees in the same place is the best way to build apiary bee stock. Just like taking the best cuttings from your vineyard, breed from your best hives. By splitting hives naturally, as old-time apiculturists used to do, rather than through the compulsory requeening of hives as some recommend (this can involve actually killing an older queen who may not be "producing" as much as a younger queen), you are working with the bees' natural instincts to raise new queens on their own. Dividing the hive to create a new hive creates the smaller nucleus colony, just like the one with which we started. By keeping the bee yard within the family, as it were, you not only promote the most well adapted of the bees, but also give them stronger resistance to typical pests that may attack the apiary. One of the most prevalent and potentially disastrous of the parasites, varroa mites, can infiltrate a hive and live off the natural life cycles of the bees. When the hive is split into a nuc for a new hive, this disrupts the bees' cycle, and as a consequence disrupts the mites' own reproductive cycle. This helps keep the mites in check because it slows down their population growth, which, when they are left to breed rampantly, can overwhelm the hive and cause the bee colony to collapse later on in the season.

There are many great resources for in-depth and thoughtful beekeeping that address all aspects of the apicultural journey, from splitting hives to thinking about swarms. When help and education are needed, ask local beekeeping mentors, or ask to lend a hand at another apiary to see how it's done. There are also a number of step-by-step books available. *Natural Beekeeping*, by Ross Conrad, is a relatively new book that is already a classic, and *A Homemade Life: Keeping Bees with Ashley English* is a good beginner's how-to. *The Beekeeper's Bible*, by Richard Jones, not only provides the essentials of beekeeping, but is a lovely almanac of history, art, and bee and honey craft.

Location, Land

When our bees arrived that Sunday and we placed the nuc provisionally under an apple tree in a part of the orchard that doesn't see much traffic, we had to make a final decision about where on the farm would we place the bee yard. Where in our landscape would suit them the best? I imagined them staying somewhere in the orchard beneath the stretching, twisting arm of an apple tree. But there are many aspects and needs of the bees' life to consider when planning an apiary. The bees must be far enough away from activity so they are not disturbed or annoyed. They must have sunshine—bees live and work in the warmth and light of the sun—but they also need some shade for those overly hot summer afternoons when the sun is merciless. The shade must never be too much, though, to account for cold winter days when some unrelenting yet milder sun would be most welcome. They need to be near a water source. Certainly, you can keep a clean bucket of water, with a bee stick (a long piece of straight branch that rests in and sticks out of the bucket, so that the bees can climb out and away from the water—otherwise they can drown), but a native source of water nearby is preferred. In open land that experiences wind, some shelter will need to be provided with a windbreak for unruly days.

VIRGIL, WHO WROTE about bees in his beautiful book on agriculture, *Georgics*, describes the ideal location for a bee yard best:

First find a site and station for
the bees

far from the way of the wind

(for wind obstructs them bringing
home whatever they may forage),

with neither ewes nearby or
butting kids

to trample down the blossom—
to say nothing of a heifer stray-
ing through the dew and flattening
growing grass.

And keep out from their grazing
grounds the lizard with its ornate
back,

the bee-eater, and other creatures
of the air, not least the fabled
swallow, Procne, her breast still
bearing stains from her bloodied
hands,

for all of these lay widespread
waste—they'll snatch your bees
on the wing

and bear them off in their mouths,
a tasty snack for greedy nestlings.

Make sure you have at hand clear
springs and pools with moss-fringed
rims, a rippling stream that rambles
through the grass, and have a palm

or outsize oleaster to cast its shadow
on the porch, so that in spring that
they so love

when sent out by the queens

first swarms of young and new bees
issued from the hive may play; a
riverbank nearby might tempt them
to retire from the heat

and, on their way and in their way,
a leafy tree entices them to tarry.

Whether water there is standing still
or flowing

lob rocks into its middle and willow
logs to lie crosswise

so they'll have stepping stones where
they can take a rest and spread their
wings to dry by the fires of the sun,

all this in case an east wind
occurred to sprinkle them while
they are dawdling,

or dunked them head first in
the drink.

Let all around be gay with evergreen
cassia, spreads of fragrant thyme

and masses of aromatic savory.
Let violet beds absorb moisture
from the rills and runnels.

On our own land, to place the bee yard in the north orchard, where the nuc was settled, would be too far from our natural source of water, the brook that flows down one side of our meadow, and the southern end of the orchard would be too close to the hum of our own work. These bees, docile Italian bees by their nature and race, can get angry like any other bees, and too much activity could be cause for alarm.

We knew they must be near water, and fresh running water is best. At first, we thought to place the apiary at the edge of what will one day be the new orchard near the brook that lines the property, the cool water skimming over moss-covered rocks in their maple and ash glade. But the bear likes the glade too, and the bee skep might be on the bear's path to visit the compost pile. I've seen him there in the morning, scrounging around amid the restaurant's leavings, rising up on his hind legs as he heard me open the door to let the cats out, then looking me straight in the eye. The risk is too high. Bears do really love honey.

A lone, very tall pine tree stands sentry at the top of the vineyard. We decided to place the hive there, on the southeast side of this tree, which would protect the beehive from wind and allow it to be warm during most of the day. For bees do truly like heat—in the Greek and Roman myths they are always creatures of the sun—but they don't like to be overheated. In the afternoon, the lone pine provides shade from the setting sun, giving the bees some time to cool down before the evening.

Even though the stream is not that far from the hive, and neither is the waterway we created between the orchard and the vineyard to divert water from the top of the field, we put a white bucket of water with the bee stick standing up and out of it so the bees can crawl up and down into their make-shift pool. Bees can travel incredibly far distances, and I'm sure they prefer the shady sylvan retreat of the brook, but the bucket is there just in case.

Choosing the location for an apiary is much like choosing the best spot for a vineyard, orchard, or garden. Every living thing on the farm has a place it prefers, where the geology, geography, flora, and fauna will help it thrive. Finding the place where the bees will live is no different. This is their terroir.

Hive, House

Usually when you purchase a nuc from a local beekeeper, you drive home with or have the hive delivered with the colony of bees. Some new beekeepers are very handy with wood,

saw, and nails and like to build their own hives, but for the novice

this is a great time to get a properly constructed hive directly

from the beekeeper. Most hives these days are Langstroth Hives,

an architecture that was established in the mid-1800s, a verti-

cally oriented hive that allows the beekeeper to easily remove

and replace frames of honeycomb thick with honey.

IN THAT ACTION ALONE, the smooth and careful motion of removing the frames from the nuc to the hive, or hive to hive, or hive to honey house, I am reminded of how long this dance has been occurring. How through the ages humans have interacted with bees. Prehistoric cave paintings dating as far back as 15,000 BC depict scenes of harvesting honey and beeswax from wild bee colonies. One of the oldest cave paintings was discovered in 1900 in Valencia, Spain, in the Cueve de le Arana—the Cave of the Spider. The image depicts an androgynous character hanging from three vines or ropes to harvest honey from a wild hive high up on a cliff. Honeybees in various sizes surround the collector, the largest bees looking as if they are quite close to us, the viewer, the smallest nearer to the figure, somehow this difference in size suggesting an early effort at perspective. The painting is done in red.

Ancient Egyptian beekeepers were at the avant-garde of beekeeping. They housed bees in stacked cylinders made of clay, dung, or woven grasses. Ancient Greece and Rome practiced beekeeping as well, sheltering bees in a variety of structures, including hollowed-out logs, pottery, and straw baskets called skeps, the form that still comes to mind when we think *beehive*. Northern Europeans first harvested honey from wild hives, using all kinds of paraphernalia for climbing the trees to get to the hives. Eventually, they too became tired of working this way, and resorted to using structures similar to those used by the Greeks and Romans.

As beautiful as these hive structures are, they have a key design flaw: They must be destroyed to get at the honey, and therefore the bees must be destroyed. It wasn't until 1852 when a pastor and educator from Pennsylvania,

Lorenzo Langstroth, a man who as a young boy wore holes in the knees of his pants watching insect life of all kinds at ground level, created a boxlike hive structure composed of interchangeable and removable frames at his home in Massachusetts. His deep understanding of what has now become known as *bee space* enabled him to create a design supremely comfortable for queen, drones, and workers alike. He had noticed that bees will create comb if they have ⅜ inch of space, and, alternatively, if there is ¼ inch of space they fill it with propolis, "a mastic made from flower blossoms," as Virgil wrote. Langstroth's design grew from what had been known as the Leaf Hive, invented in Switzerland in 1789 by Francois Huber, a Swiss naturalist—son of a famous theological writer and an honored military man with an interest in ornithology—who at the age of fifteen developed a disease that resulted in blindness. With the help of his wife, Marie Aimée, and his servant, Francois was able to carry out investigations that laid the foundation of our scientific knowledge concerning the life history of the honey bee.

In 1852, Lorenzo Langstroth filed a patent for his new hive design, and by the end of the year he had over a hundred of them ready to sell. He spent many years defending his patent, but never could successfully. The Langstroth Hive has remained, essentially unaltered, for 150 years.

Another hive design has recently become very popular over the last several years, along with a renewed interest in beekeeping: the Top Bar Hive. But before considering the advantages or disadvantages of the Top Bar Hive, understanding the basics of the most recent standard hive, the Langstroth, is an important part of the decision making that goes into how to house bees.

Most modern hives are made with wood, usually pine or cedar. Pine is the best, as it is relatively neutral in terms of scent. Cedar is too fragrant for bees. They find the smell confusing, and bees in cedar hives spend a lot of energy coating the inside of their hive with propolis to seal off the odor. An organically minded beekeeper would never consider chemically treated wood for the hive, and in organic certification it is not allowed. Wood, metal, and wax are the traditional components of the traditional hive, though now plastic, polystyrene, and Styrofoam are making appearances, especially in larger or traveling apiaries. Plastic frames are becoming quite popular, and they do have a number of pluses in terms of economy and ease of use. They are sturdier than beeswax, so they break and crack much less. They are also more temperature-resistant, they do not become brittle in deep-cold weather, and they do not easily succumb to the vagaries of mice and other pests. If you are building your own frames, plastic ones take less time to put together. They are certainly less expensive than buying beeswax combs or using the beeswax combs from within your own bee yard. Employing your own beeswax comb sacrifices a portion of your honey-producing square footage for a period of time while the new hive gets up and running.

Even though there may be a lot to recommend the plastic frames, and as much as beekeepers might prefer them for their facility and economy, the bees themselves do not. I have heard it said that generally bees will not draw out comb (or make honeycomb ready for pollen deposits) on the plastic frame unless it is already coated with beeswax or sprayed with sugar water. They are finicky. And rightly so.

Why bees might not like a plastic frame seems rather obvious, the most blatant reason being that the plastic frame is not what they naturally would set honey on; it is not part of their natural, instinctual process. I'm sure it doesn't feel the same, smell the same, or taste the same to them. There are also theories that the plastic may emit minor amounts of chemicals that disturb the bees in some way, or the plastic is not permeable enough for the signals and gestures, the bee dance, they perform to effectively communicate with one another on both sides of the comb.

This is one of the parts of the story where I wish we had done further preparation before our bees arrived, because our nuc and our hive came with this acceptable but not preferable black plastic faux-comb foundation. Any new hives we start from now on will want beeswax, and we will replace these curent plastic frames in the hive as we can.

If the usual issues with plastic were not enough to promote the use of beeswax—the potential chemical issues as well as the issue of how to properly discard the frames when they're no longer useful—the knowledge that poly-styrene has been encouraged for our northern climate as well, because it is a material that has good insulating properties, is disconcerting. Polystyrene is a petroleum product that emits off-gassing chemicals. The insulation that is touted for our cold weather can turn against the bees on a sunny, midwinter day. The insulation works two ways: keeping cold out on cold days, but also keeping heat out on warm days.

Wood and beeswax are the best, most natural materials to make bees happy in their new home. But wood will eventually rot out there in the elements, especially a soft wood like pine. To help preserve the hive, paint it with latex-based paint. Using color can help the bees identify their home from far away. Oil-based paints contain too many toxins and they seal the wood too well, promoting cracking. Bees generate moisture and heat, so the wood needs to breathe, allowing moisture to pass through. Latex paint has breathability, plus its glossy texture will help repel rain. Painting only occurs on the outside of the hive. Bees only want bare, untreated wood and beeswax on the inside, which they then coat with a fine layer of propolis.

Some large-scale beekeepers dip the wood for the hive in wax with preservatives to protect the hives, and from a glance this may seem like a natural and nifty way to go, but those wood preservatives are often petro-

leum-based, and therefore are toxic to the bees. Linseed oil can be a good alternative to latex paint and can be mixed with a bit of turpentine to help it dry faster and soak deeper into the wood. It's important to check the ingredients on the turpentine. It too can have petroleum-based chemicals in it, which can cause havoc for the bees.

Some biodynamic beekeepers paint the outside of the hives with a mixture made from propolis and food-grade grain alcohol, simulating what the bees do in the interior of the hive. This may very well be the healthiest and most natural way to approach preserving a hive, but it presumes a significant amount of propolis with which to work and to remove from the hives. Leaving propolis in the inside of the hive is key—it is what helps the bees stay healthy and happy, their own internal medicine.

Since the popularity of the Top Bar Hive method has risen so dramatically, there have become two somewhat adversarial camps: one that maintains that the Top Bar Hive is more natural and better suited to bees, and the Langstroth Hive camp, which says simply that this isn't so. Like all confrontational conversations, the sides sometimes become a bit blind, with proponents of either method refusing to see the benefits and purposes of the other.

The Top Bar Hive was created by a research team trying to adapt movable frames for Kenyan hollow log hives that were often hung from a branch in a tree. These are African hives with Africanized bees. The Top Bar Hive was developed in a trapezoidal or triangular shape with slanted sides, though there is no specific set design. The bees are encouraged to make their beeswax foundation moving side-to-side rather than vertically as in the Langstroth Hive. There is no foundation and no spacing for the bees; they must do it all on their own, just as they would in a natural environment. The Top Bar Hive is also terrifically inexpensive to make, requiring minimal tools and resources. This is perfect for the tribal circumstances in Africa where wood is in short supply and sophisticated tools are hard to come by if not nonexistent. You can make a very expensive Top Bar Hive if you want, as there are no current rules about how to build them; when trying to do it in the simplest way possible, the cost is significantly less than fabricating the more exacting Langstroth Hive. Top Bar Hives are much easier to build, without the strict requirement of measuring bee space as in the Langstroth Hive, so for the beginning beekeeper who wants to construct his or her own hives, this is all very appealing.

But the many advantages of the Top Bar Hive, even its naturalness, can also be accomplished in the Langstroth Hive. The biggest argument in favor of the Top Bar Hive is how the bees create their own comb. However, this can be done in the Langstroth Hive as well with just the starter beeswax combs set at the top or the sides of the frame. We've also learned from mentors that the European bees that typically get raised here in New England like to work vertically and will work vertically when left to their own devices in the wild; they prefer to make their hives in the hollow of a tree and work downward rather than horizontally. Given this observation, the Top Bar Hive doesn't seem as advantageous for our bees.

The Top Bar Hive was really created for the Africanized bee, which is very aggressive. Given that the Top Bar Hive method doesn't provide any visible bee space from the top, the hive is less disturbed when it is opened and closed. This is rather brilliant when working with aggressive Africanized stock, as the design keeps the hive calmer, and, while disturbing the bees as little as possible is always preferable, European bees are much gentler and don't have the same responses as their Africanized cousins. However, this Top Bar Hive attribute does not offer the European bee enough of the same advantages to make it better to shift from Langstroth to Top Bar and actually is less ideal because of how the design forces the bees to build comb.

The kind of hive you have and the way you preserve it and manage it may seem like decisions that should be up to each individual beekeeper, and, certainly, each beekeeper should create a bee yard, no matter how small, that works well with his skill level. But ultimately, just as in any kind of farming, the beekeeper should attempt to understand and observe his bees and how they want to work naturally. Then you can set up the bees' hive in a way that makes the bees most comfortable and healthy.

Api Culture

Coenobium, or *cenobium*, is a Latin term that combines the Greek *koinos* or "common" and *bios*, "life." Coenobium also describes a colony with a fixed number of cells. Coenobium is used to refer to monastic communities where monks or nuns look to live in communal spiritual endeavor. It is also the name of one of my favorite wines made by two cloistered Cistercian nuns. Or it can be used to describe the working of a beehive, the hexagonal structure of each cell joining the next, the bees living in a rigid communal social order, the beehive a nucleus (the word no misnomer here) in a larger structure, the center of the bees' domain, and the beating, buzzing, golden heart of the natural world.

THE LIFE OF THE HIVE EXAMINED is nothing short of miraculous. Correlations have long been made between the cooperative nature of work, life, and death within the hive and our own human structures of living. There is much to be gained in both a practical and philosophical study of the bees' life. And while we share in their honey, we can also share in the lessons they teach us about the life of the individual in community. Bees don't live in a vacuum and neither do we.

On the practical level of caring for bees, understanding something about their social habits and concerns becomes rather necessary. In their regimented structure, everything begins with the queen. To the unguarded eye, she is a little bit larger than the others, a commanding presence. Her body is elongated, with shorter wings and a shorter thorax. Drones and worker bees are different yet again. The drones, or male honeybees, have large compound eyes, bodies that are longer than the worker bees' but still shorter than the queen's, and rounder bottoms. Drones lack stingers, the furry pollen baskets attached to a worker bee's legs, and wax glands. They have one purpose, and one purpose alone: They are alive to mate with the queen. They make occasional flights out into the sky, looking for a queen in midair with whom to mate. Otherwise, they are cared for by their little worker bee sisters: cleaned, fed, and pampered. Not a bad lifestyle, you might say, but in reality the drones have somewhat of a raw deal. They have no stingers with which to protect themselves, and after they mate, they die. Their genitalia stays attached to and flies away with the queen.

The female worker bees are responsible for all the activities in the hive: nursing, attending, cleaning the hive, cleaning other bees, undertaking (removing dead bees from the hive), building, capping, packing, ripening, repairing. They are designed to be much smaller than the queen or the drones, they have broad scent and food glands, and they are the ones with those furry little pollen baskets on their legs. In the short lifespan of the worker, their days are divided into different activities depending on their age. As she grows up, the worker bee's wax glands and stinger develop. She works either from home or in the field depending on where she is in her life cycle. If you encounter a bee out in the field collecting pollen, you'll know she is older in age. A worker bee born in summer lives for about six weeks, working all the while. A worker bee born in the fall may live up to six months, residing in and maintaining the hive during the winter months.

A summer worker bee begins working outside after three weeks. Then her duties shift to collecting pollen, propolis, and water and guarding the hive from intruders, using her stinger or releasing a pheromone that will alert her hive mates and sound the alarm.

In working with bees, there are a few essential items needed: a smoker, protective clothing, and a hive tool. Because the bees can become alarmed and disgruntled, even the gentlest bees, providing a calming influence makes for a happier visit to the bee yard. Smoke is one of the oldest tools, one that beekeepers have used for centuries. The smoke confuses, distracts, and disarms the bees living in the hive. The smoke actually confounds their communication methods.

Animals and insects fear fire in a way that we as humans do not. We see fire as a beneficial tool: We've captured it for heat and for cooking. Animals and insects have never harnessed flame or understood it in that way. If they smell smoke, they must assume that fire is not far behind, and their experience of fire usually ends in destruction.

The smoke the beekeeper uses sends a note of alarm throughout the hive, and the worker bees begin to gorge on the honey in the comb, preparing to vacate the hive because they believe a fire will soon arrive to destroy their home. When the bees are distracted by imminent disaster, they are not so concerned with the hive being opened and examined. In addition, the clever smoke masks the scent of the pheromones given off by the bees guarding the openings to the hive—a biochemical signal they give off in order to organize their defenses with the other bees in the hive. The effect of the smoke breaks down all communication.

When I first learned about the use of and effects of the smoke on the hives, frankly, it made me uncomfortable. The thought of stressing the bees to such an extent seems unfair and cruel. Then we take honey from them, which only adds to their humiliation. But the reality is that if we want to

FIRING THE SMOKER

raise bees to share in their honey, we need to be able to interact with them in a way that harms both us and them the least. We must treat them fairly. A little bit of smoke goes a long way. When we harvest honey, we should never take the whole supply. Beekeepers who remove all the season's honey condemn the colony to death, as the bees will not be able to replenish their stores for the winter. Such a beekeeper may think that reaping all the honey and buying a replacement package of bees is more cost-effective. It is also myopic and selfish. That same beekeeper who cleans out the honey in the hive may argue that she can maintain the hive by feeding sugar syrup in the winter, but bees don't want to really eat sugar syrup—they only do so to survive—and it certainly isn't a practice that promotes honey as expression of terroir. Honey made from the wildflower pollen surrounding a beehive's landscape says something of that landscape, whereas honey produced from sugar syrup says nothing, or at any rate nothing particular.

The smoker used by beekeepers has a classic shape, looking a little like the Tin Man from *The Wizard of Oz*. Since smoke for distracting bees has been used for millennia, the form of smoke delivery has taken a variety of shapes. A smoldering stick is thought to have been one of the earliest forms of smoking the hive, or a pan of spent coals from the fire. Bellows were adapted from smiths. In the mid-1800s, a gentleman named Moses Quinby, one of the first commercial beekeepers in America, apparently designed the modern hand-held smoker that is in use now, a bellows attached to a tin burner.

The smoker is fairly self-explanatory to use: You make a small fire in the bottom, get it going, close up the canister, and use the bellows to blow smoke. The important note here is to consider what you are burning. Bees are highly sensitive to toxins, and smoke, by its very nature, already has some level of toxicity. Neutral kindling substances, like pine, pine needles, and dried grass are all good to use. You can even use a fresh, green grass on the top of your fire to help cool the flames in the tin. Otherwise the fire can get carried away

and start to spark and spit flame around the bees, and this will upset them rather than calming them. Do not use newspaper, because the inks used, even soy-based inks, are too toxic for the bees, and don't use burlap, though it is often recommended, because it is coated with an oily substance that carries a heavy scent, burning or not.

Wearing protective clothing when visiting hives is also wise, especially as a beginning beekeeper. Having that added protection helps keep the beekeeper calm around the bees, and this is one of the most important and often undiscussed tools of beekeeper. Your own mental state can greatly affect the bees. Visit them when relaxed, focused, and in a good mood. Protective gear, like the classic bee suit and veil, gloves, and elastic-bottomed pants down around tall boots, not only really does the job of protecting against potential bee stings, but also gives the added comfort of knowing you are protected, taking away the sting of fear. Also, never attend to bees when angry. They will sense that immediately, and it will rile them as well.

Working with bees in the evening when they have returned to the hive for the night and are quieting down can make for an easier visit. They will already be naturally calmed by the waning sun and coolness of sundown.

The hive tool is a necessary piece of equipment used to help remove frames from the hive. You can buy a fancy Italian one at a beekeeping supply store; our friend Dean suggested we go to our local hardware store and purchase a miniature pry-bar with sharpened blades for about a quarter of the price.

There are many experienced beekeepers who can work with their hives without protective clothing and without smoke. They have a particularly serene demeanor that keeps the bees calm, and they spend time with their bees, so the bees know them and trust them. Many natural beekeepers also promote leaving the use of smoke behind because the reality is that smoke is not good for bees and it's not good for us to breathe in either.

But for the beginning beekeeper this advice would be naïve, and refusing to use smoke or protective clothing could result in a lot of beestings and a lot of dead bees, which is sustainable to no one. When the bees sting you, their stingers can get left behind in your skin. Once the stinger is removed from a worker bee, it dies. It's much wiser and more humane to work with a little bit of smoke to help manage the bees until you have garnered so much experience you can choose otherwise. Keep in mind that those beekeepers who successfully work with no smoke are few and far between.

It is an amazing gift to go out to the apiary and open the hive with the smoke wafting around and with the bees already very quiet because the sun has already gone behind the western hill. When you lift out a frame, you can see a bit of the comb that they have already started to form, that waxy hexagonal covering that they create by shaping the six-sided cells with their mandibles. The bees move and dance on the framework, and you watch them in silence. You are quiet because you must be quiet. Not just because to be loud would upset the bees, but because it is a humbling experience to watch them work. Small and wondrous creatures cloaked in mythology, they are nature writ large on these panels of honeycomb. They are an intuitive collective that allows us to participate in their rituals and society, to partake of the honey they create, this life-giving substance whose process is also life giving. This is wildness sharing in its domesticity and showing us the basic foundations of our own existence. I believe it was Einstein who said, "If honeybees were to disappear, humans would have only four years to live." He may not have been entirely accurate in this statement, but his point is well taken. The primary work of the honeybee, the collection and translation of pollen into

THE DISAPPEARANCE OF BEES

Any discussion of bees today must include some information on Colony Collapse Disorder. Colony Collapse Disorder (CCD) is defined by the unprecedented disappearance of the worker bee force of a honey beehive. The queen remains, honey remains, and unborn bees, or capped bees, can remain. What is crucial about these signs is that the queen is left behind in the hive. It stands to reason that if the hive loses its queen it will disintegrate socially, and worker bees will die or leave to find another queen, but for the worker bees to leave the queen behind? Anarchy.

As early as 1900, evidence of CCD had been documented, particularly in 1918 and 1919. Initially described as the mystery disease, it eventually became known as the disappearing disease. From 1972 to 2006, dramatic declines continued among wild honeybee populations in the United States (my wild hive experience was in 1969), and since then there has also been a gradual decline in the number of colonies maintained by beekeepers. The reductions in the numbers of hives were attributed to a variety of reasons: urbanization, pesticide use, tracheal and varroa mites, and a generation of beekeep-

ers retiring. But in 2006–07, this decline dropped precipitously to new lows, and the phenomenon became known as Colony Collapse Disorder in an effort to raise our own alarm and to describe this sudden and more extreme rash of disappearances.

The first report of CCD came in November of 2006, from a mid-Atlantic beekeeper overwintering in Florida. By February 2007, migratory beekeepers reported 30 to 90 percent losses of their colonies, and some colonies were completely obliterated. By late February of that same year, non-migratory beekeepers in the mid-Atlantic and Northwest were reporting 50 percent losses. Numbers all too high to grasp.

In July of 2009 the first annual report of the US Colony Collapse Disorder Steering Committee was published. By the second annual report, the committee found that no single factor seemed to contribute to the phenomenon, but rather there was a combination of factors including toxic pathogens, pesticides, and parasites. However, it was noted that there were no damaging populations of varroa mites or the nosema parasite found in colonies that had suffered—one of the go-to general theories behind this decline.

What they did find were sublethal effects of some pesticides, including pesticides registered for use by beekeepers for controlling varroa mites. Sublethal effects were also found from pesticides in the neonicotinoid and fungicide families, pesticides that researchers believe adversely affect the nervous and immune systems of the honeybee. Sublethal means almost lethal, almost fatal—not killing outright, but causing a physical havoc worthy of note.

In a large study done in 2010, pesticides were also found in the beeswax and bee pollen of affected hives, and links were confirmed between poor colony health, inadequate diet, and long-distance transportation.

Industrial beekeepers spend all year traveling with their hives from large monocrop to large monocrop for pollination purposes. But one of the problems with monocrops is that, once the crop is done flowering, there is nothing else for the bees to eat. There is no diversity of flora or fauna. So the bees are transported to the next blossoming crop, and fed high-fructose corn syrup and flower pollen from China on the way to simply keep them alive. As Michael Pollan, author of *The Omnivore's Dil-emma*, says in the provocative and melancholy film *Queen of the Sun*, a documentary about the plight of the honeybees, "Nothing is more viscerally offensive than feeding the creators of honey high-fructose corn syrup."

In 2012, two separate studies published in *Science* magazine cited that neonicotinoids (a class of insecticides) may interfere with the bees' natural homing abilities, causing them to become disoriented and unable to find their way back to the hive.

According to the US Department of Agriculture, there has long been concern that pesticides and some fungicides may have sublethal effects on bees, not killing them immediately, but instead impairing their development and behavior, and of special interest is this class called neonicotinoids, which contain as their active ingredient a chemical called imidacloprid, as well as similar other chemicals. Honeybees may be mostly affected when these pesticides are used as seed treatments, because when used this way, the pesticides are known to work their way up through the plant and into the flowers and nectar. Scientists have noted that these residues taken up by bees may not be immediately lethal, but they are concerned by chronic long-term problems indicated by overexposure.

Genetically engineered Bt corn is treated with neonicotinoids. In a 2012 study, scientists found that the insecticide was present in the unplanted soil of neighboring farms next to those planted with Bt corn as well as in dandelions growing nearby. Studies by the National Institute of Bee Keeping in Bologna, Italy, suggest that large amounts of neonicotinoids used as a preplanting treatment for the seeds of corn and sunflowers were carried back to honeybee hives. Another Italian study found that the pneumatic drilling machines used to plant corn seeds dressed with neonicotinoids release large amounts of the insecticide into the air, causing significant mortality in foraging honeybees. So much for "sublethal."

A recent in situ study here in the United States provided strong evidence of exposure to sublethal levels of neonicotinoids in the high-fructose corn syrup (made from Bt corn) being used to feed industrial honeybees when foraging, and suggested that this causes bees to exhibit symptoms of CCD twenty-three weeks after ingestion.

And what is Bt corn you might ask? Bt corn is a genetically modified organism (GMO). To transform a plant into a GMO plant, the gene that produces a desirable trait is identified and separated from what is known as the donor organism, which might be a bacterium, a fungus, or even another plant. In the case of Bt corn, the donor is a naturally occurring soil bacterium (*Bacillus thuringiensis*), and the gene of interest produces a toxin that kills the European corn borer. The GMO only works if the insect ingests a part of the plant containing the Bt (it is important to note that not all parts of the plant contain equal amounts of the Bt at all times, nor are all the targeted larvae equally susceptible).

Around 90 percent of all corn seeds planted in the United States are coated with neonicotinoids and with such a high percentage of those being Bt corn seeds, one might wonder why they must be coated with an insecticide in addition to being genetically modified. According to GMO production companies, one of the "sustainable" reasons for genetic modification is supposedly to enable the farmer to limit the use of insecticides and pesticides.

In relation to the honeybees, it seems clear from numerous reports that a variety of factors related to insecticides and resultant immune deficiencies, behavioral degradation, and ultimately death are responsible for what we've called Colony Collapse Disorder. The skeptic cannot deny it: Our honeybees are disappearing. And if the honeybees are disappearing, what of the rest of our landscape? What of us?

honey, helps produce 40 percent of the food we eat, and in turn affects other animals: the birds in the forests and meadows, the livestock that graze, the smaller insects and animalia that we can't even imagine. And whether we can imagine this or not, the fact remains: Honeybees are the keystone to l ife as we know it on this planet. Without them, the structure of our world breaks down.

Seasonal Rhythms of Beekeeping

O nce bees are successfully ensconced in the bee yard, there are seasonal aspects to working with the bees. In spring, summer, autumn, and winter a series of chores goes a long way toward keeping the bees happy and in fine fettle. Because we have not spent many seasons yet with our bees, I cannot offer the long, intricate list of tasks a passionate and wise beekeeper might move through each season, but I can point out the basic rhythms of the seasons with bees.

IN EARLY SPRING, when the temperatures fluctuate between warm days and cold nights, or hot days and then an unexpected snow, bees are at their most susceptible. The warmth from the new spring sun and the longer days encourage the queen to lay eggs. If the brood grows fast during this time, the hive can very quickly eat through the remainder of their winter food stores, leaving them starving before the flora catches up to their activity and provides blossom and nectar. The beekeeper's task during this period is to watch for such diminishing stores, potential swarming, and vandalism from the also waking world of animal predators. If the bees were left with enough honey from the previous summer, they should be able to get through the winter months easily, but each season and each hive is different.

On those warm, sunny spring days, when the temperatures start to get up in the 50s, the time to look inside the hive has arrived, checking for life, honey, the queen, and new eggs. Discovering a dead hive is a sad moment, but it also provides lessons, an opportunity to learn how to care for the hive better. If the honey is not infected by a virus, it can be offered to other hives in need of supplemental food, which is better than some of the other feeding options.

If the hive is thriving, but the food stores look low, you can do some of that supplemental feeding to get them through the thin days until the new spring plants can provide nectar. Feeding bees at the end of the winter and in early spring offers a bit of controversy, but only in what you feed them. If you don't have honey from a dead hive, or another source for sealed honeycomb, extracted honey from a known and pathogen-free source can work well. It is always better to feed bees unextracted honey, rather than feeding them honey already harvested, but each situation necessitates what is available to offer. The best food for bees is from unheated honey or what is called bee bread. A standard practice for making a honey syrup for bees is to add a bit of honey to a bit of water and boil it to purify it. But, in reality, there are some chemical changes that happen to the honey during this process that aren't as good for the bees. It has been found that these chemical changes produce a compound that is actually harmful to the bees.

Not only is the serving of boiled honey becoming controversial, but so is the commercial practice of feeding bees in need a diet of high-fructose corn syrup. It's sweeter and more economical than using sugar, but loaded with pesticides and genetic modifications that can destroy a colony. However, many large commercial and traveling beekeepers feed their bees in this way, and not just at that crucial juncture of winter into spring.

Contrary to what a lot of new beekeepers think, if you can't feed your bees raw unheated honey, it is better to make them a sugar syrup rather than boiled honey. Even more counterintuitive is that it is better to use white sugar than to use brown or raw sugar. The latter contain elements that are difficult for bees to digest. Be sure the sugar is from sugar cane and not beet sugar. Around 90 percent of beet sugar is made from GMO beets.

While sugar syrup can get the bees through a tight spot, I can't stress enough that honeybees are only meant to eat honey, and a sugar syrup is an option devoid of all the benefits in calories, minerals, and vitamins that honey gives to the bees. The sugar syrup is essentially something to resort to just to keep the bees alive for a very short time period.

One way to help mitigate the lack of health benefits of the sugar syrup is to follow a biodynamic protocol and make a "bee tea." While it's still important not to rely on the tea too heavily, it can be more helpful than a straight sugar syrup if the bees need sustenance between winter and spring, or if the hive has been robbed, or if you are starting a new hive from package bees and you have no honey stored in a hive. Most sugar syrup recipes call for a one-to-one sugar-to-water mixture in the spring. The biodynamic preparation we use is a little more than two-to-one sugar to water with the addition of dried chamomile and/or thyme with a little sea salt added in.

By the time the spring gets warm enough, we break down whatever we have used for winter insulation and begin weekly visits to the hive checking for swarm cells. Spring "cleaning" involves removing old frames and replacing them with new frames, and updating any equipment needed. Splitting the hives would occur right around now, as would the natural supersedure (replacement) of queens.

In midspring, we continue to monitor for a swarm and make a plan about how we will capture and use a swarm should one occur. As a beginning beekeeper, it's a good idea to be in conversation with a beekeeping mentor about what to do regarding a swarm and how to implement a plan of action. To help mitigate a swarm, you can reverse the hive bodies, placing the top hive body, where the bees will have clustered for the winter to keep warm, on the bottom, and putting the relatively empty bottom box on top. The bees won't work downward to look for new space, but they will work upward. Putting the empty hive body on top gives them a place to grow into.

If the spring proves pollen-intense and the bees go into full honey production, there may even be a late spring honey harvest. At this time, it's critical to keep an eye out for any problems like varroa mites.

The work of a beekeeper in summer slows down significantly. The heat and nectar flow of the height of summer will see the bees working at their optimum. The warmest part of the summer will be the best time for harvesting

BEE TEA

The recipe for 1 gallon of tea is as follows:

16 cups white cane sugar

1 teaspoon natural sea salt

6 cups hot tap water

2 teaspoons fresh or dried chamomile and/or thyme

Combine the sugar and salt. Add the hot tap water to the mixture and stir well. Don't boil the sugar and water, as that may make the sugar caramel-ize. Separately, do boil another 2 cups of water and steep the herbs, covered, for about 10 minutes. Strain that tea, then mix it into the sugar-and-water concoction. Mix thoroughly until the sugar is completely dissolved.

At this juncture, many a beekeeper would allow the tea to steep, then fill their bee feeders to take out to the bees. We don't have a feeder since it is usually too cold to set it up when feeding might be needed. We spread the sugar syrup tea out on a cookie sheet and let it harden, then place the pieces in the hive, centered over where the colony is clustered.

Adding the chamomile and thyme to the sugar syrup provides an immune boost to the colony, giving the bees something more than just the sweet filler of the sugar alone. If the honey stores look low in late summer or early fall, it is best to feed the bees the tea with a feeder at that time in order to avoid having to feed them in the midwinter when temperatures are cold and the bees may not be able to reach the food source.

POTTING BEE-FRIENDLY FLOWERS

the honey, unless there is a lot of rain or a drought that affects the ability of the bees to collect pollen. In either of those situations, it's important to leave them their honey for food. If honey has already been extracted and you hit a dry or wet spell, the beekeeper needs to keep an eye on the honey stores to determine whether the bees need to be fed. And, as always, if the conditions are conducive to harvest, leave the bees enough honey to get through the upcoming winter.

Even though the work becomes lighter, we still visit with our bees regularly to check on the queen, honey, brood cells, and general activity of the hive. You have to always be on the lookout for that potential swarm. At this time, we add more supers as needed, or even a second brood box. If the hive is overflowing with bees, they are communicating that the queen needs additional space. We continually look for signs of disease and varroa mite populations. If there are unwanted pathogens, it's good to be in contact with a mentor to discuss the best options for how to deal with these issues naturally.

By the time autumn arrives, the queen will start to slow down her egg laying and the pollen sources will start to die down as well. The bees will start to prepare for hibernation, working extra hard to protect and coat the hive with propolis. At this time, we space out our hive visits more. Once every two weeks suffices. We check for the queen, eggs, capped and uncapped

brood, and a healthy laying pattern. We take a good look at capped honey storage. This is the point in the season for assessing the need for autumn feeding of the bees. If necessary, providing them with food early enough in the fall allows them time to convert the bee tea to honey for the winter.

By the end of the fall, open hive visits should stop. You don't want to expose the interior of the hive to frigid temperatures.

Late autumn will be time to insulate the hive for the colder winter months. Many beekeepers use black roofing paper, wrapping the hives in this winter coat. We have used rigid foam insulation cut and affixed snugly to the sides and top of the hive. We also put a rock on top of the hive to keep the lid on when brutal winter winds come to call. On a warm winter day with loads of sun, we'll remove the top layer of insulation so that the heat can get into the heart of the hive.

The nice thing about using tar paper or roofing paper is that the black absorbs the sun so much on a sunny day that it can warm the hive up enough that the typical cold-climate cluster the bees form around the queen can disband for the day and get to the honey stores. Sometimes, in very cold winters, the bees may have plenty of honey, but it is too cold for them to separate from the cluster, and they can starve because they just can't get to the honey source.

Having said all this, we have since learned that well-respected natural beekeepers suggest that wrapping or insulating the whole hive in a cold climate isn't necessary, or at least not in a cold climate like Vermont. (They will concede that if you have consistent below-zero temperatures, insulation is needed.) In a healthy hive with good winter honey stores and a dry interior, the bees will happily survive through the winter. The biggest trick is keeping the moisture out with all the inclement weather and the fluctuating temperatures that occur in Vermont from January through April.

The bottom of the hive can be vulnerable to moisture because of rain and snow getting in on the ground floor. Canting the hive a bit forward by propping it in the back with a board can help any unwanted water drain away from the hive. Bases of hives that have a screen don't need to worry about this, as the water will drain right through.

Most of the moisture in a winter hive is caused by the bees themselves, who generate heat and condensation from their own work at keeping themselves and the queen fed and warm. Hive ventilation is crucial in the winter, though it seems to be opposed to buttoning up the hive and helping the bees stay warm throughout the winter months. Many beekeepers and manuals will recommend entrance reducers for the bottom of the hive to help retain the warmth created by the bees in the winter. But the reality is that such a reduction in airflow traps moisture in the hive and causes more harm than good. This is the concern about insulating the hive as well—that too much of a good intention can backfire. We have even been advised to add ventilation to the hive by drilling a ¾-inch hole in the hive body just below the handle. This provides an additional entrance that also improves ventilation. Especially in winter, when the bottom entrance can get blocked by snow or dead bees. Regardless of whether you add additional entrances and ventilation, after a big snow you still need to dig the hive out and keep it away from the moisture of the snowpack.

In a northern climate like ours, the location of the bee yard holds as much importance, if not more, as the idea of insulating the hive. Windbreak from a tree and direct sunlight go a long way toward preserving the winter hive environment.

A great way to help keep the bees warm and dry in the winter is to insulate the top cover rather than the whole hive. On the inside of the cover you can use a piece of 1-inch-thick rigid insulation foam between the inner and outer covers. I've heard of beekeepers using loose straw or dried leaves as well. This material keeps the heat created by the hive from leaving the building, just like good roof insulation does on our houses. In addition, and perhaps even more crucially, it keeps the moisture the hive makes from collecting and condensing at the top of the hive, freezing when it gets brutally cold, and thawing and dripping during a thaw. Bees can stand to get a little wet when it's warm, but when they are cold and wet, winter life in the hive will not end well.

While bears are the New England beekeeper's fear in the summer, mice can be a problem in the winter. When temperatures get cold, a cozy, dry hive with bees congregating in one specific area and distracted by their concentration on keeping warm can be very seductive to the average field mouse. They can easily move into the bottom section of the hive to make a nest. They don't necessarily bother the bees, as bees and mice don't like to assemble together, but mice can cause damage to the comb and the frame, chewing and spitting while building their winter home. And mouse droppings and urine can make the hive an unsanitary spot for the bees. If the bees are healthy and strong, these things won't greatly affect the hive in the end, as once the hive wakes up in the spring and the temperatures get warmer, the mice will move out. But if the bees are weakened and die off over the winter, the mice can take over the whole hive, contaminating it for future use.

Applying a mouse guard or excluder can be a good late fall precaution. Narrowing the size of the hive openings so that the bees can still go in and out, but the mice cannot get in, will go a long way toward preventing a mouse in the house. Bee supply companies often sell mouse excluders made from a piece of wood with holes drilled into it that can be slid into the bottom opening of the hive. The holes are just big enough for the bees, but too small for the mice. The drawback with these kinds of excluders is that they reduce the needed airflow for the hive to stay dry during the tougher winter period.

A piece of what is called ½-inch hardware cloth, a metal mesh that is easily bendable by hand, can be cut to fit into the main opening of the hive. Cut the metal cloth to the width of the hive and 3 inches tall, bending the bottom of the mesh at a 45-degree angle so it fits snugly into the horizontal space. This handmade guard will work very well at keeping the mice out and the bees and air free flowing. Putting in the mouse guard before the temperatures get cool is key. If the excluder goes in too late, there may already be a mouse in residence, and instead of being kept out, the mouse is trapped within.

Even though life in the hive is dormant during the winter, we are still aware of what is happening with our bees. We go out periodically to tap on the hive and listen at the side of the hive to make sure we still hear their rather incessant hum. After every storm, we check on the entrance and push snow away. By gently picking up the hive, rather than opening it up, we can gauge by feel how heavy the honey supply might be. On warm winter days, even when it seems cold, some bees will still venture out. Fortunately the brook at the edge of the meadow is always running at full throttle during the winter, so they have a good fresh water source, but it's a good idea to still keep that bucket with bee stick close to the hive. Changing it out regularly if the weather remains cold is necessary, as it will freeze over quickly and become useless.

Biodynamics and Bees

Just like all the other aspects of our farm, we approach our beekeeping from a biodynamic perspective and practice. Though if I were to be precise, I would probably not use the word *beekeeper*, as it connotes the idea that we can actually "keep" bees for ourselves. Not unlike farmers who "have" bees. We cannot keep bees; we cannot have bees. They are unto themselves, and it is they who deign to share with us. Most biodynamic practitioners with bees would consider themselves bee stewards.

THE POINT HERE ABOUT SPLITTING HAIRS in a name is not unlike raising the differences in the words *winemaker* versus *winegrower* or *wild-gatherer* versus *forager*. I usually don't consider myself a winemaker; I think of myself as a winegrower because the work I try to do is in the field, not in the cellar. Any craft or art in the translation of our fruit into wine happens in the nature of the season and my response in care during that season. By the time we harvest the fruit, my role is to pick the fruit when I believe it is ready to become the wine it wants to become, and to provide a clean space and appropriate conditions for the fermentation. I try to respond to the essences the fruit puts forth in any decisions I might make about how to ferment it. This wine might ferment in an open vat on its skins for six weeks, and this other wine might get pressed immediately and ferment in a 14-gallon glass demijohn. Another wine might ferment for a while in a glass demijohn, then finish its first fermentation in a bottle to become a *pétillant naturel*. Certainly, my decisions affect the wine, and that is my part of the terroir equation. But I in no way enforce my desires on the wine. I try to work in opposition to this; I try to respond to the suggestion of the fruit, not unlike a sculptor approaching a piece of stone and trying to intuit the form the stone wants to take.

The biodynamic bee steward approaches the hive in the same way with the same adage: Stay out of the way, and make a clean environment for the bee yard. Do not bring constant outside inputs via queens and colonies or queens artificially inseminated in the lab and brought in from elsewhere. Do allow swarms, colony splitting, mating, and the bees' natural choice of their queens so they will thrive in the apiary. Leave the bees as much as possible to their natural inclinations; they will correct their own imbalances if provided with a clean and healthy environment. They know far more than we do about what is right for them. And if loss occurs, a biodynamic bee steward must understand and accept that this is part of the natural selection of the natural world.

I remember the first few nights the bees were with us and how we closed the wire mesh door to their yard once the dusk had settled, to keep them safe from any roaming creature, and to help them understand that this was now their home. We noticed every evening as twilight came, they returned to their makeshift hive like children being called to dinner. I am reminded of being a child myself on such twilit evenings, the air thick with humidity, but dusk providing something cool like a freshly laundered sheet in a north-facing room. I think of the jam jars, lids with small holes poked in them that we'd take out into the horses' meadow and catch fireflies to make lanterns full of glowing life.

HAVING THE BEES on the farm surrounds us with their magic and activity. They pollinate the orchard, the roses, the vegetables, and even though grapes are self-pollinating, the bees work in the vineyard too, participating in the creation of the unique diversity of fruit and flavors the wine from this place offers.

They remind us to be humble in the face of the wilderness, in the face of nature, and that our abilities are not really our own. We grow a tasty carrot, a unique bottle of wine, a pretty and succulent apple; we harvest delicious and fragrant honey, not because of our own will and effort at shaping what this farm can offer, but because we must work in concert with the natural inclinations of our landscape.

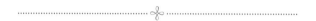

Wild Wood

When you come to visit at the farm, after walking through the vineyard and the orchard and the gardens and the greenhouse and meeting the bees, it's lovely to walk the perimeter of the meadow. I like to walk down behind the greenhouse and the old compost pile that we once tried to dress up with a dark gray painted and slatted fence that was one of my first carpentry projects.

NOW THE OLD COMPOST IS HOME is home to a wild thicket of roses and raspberries, true Queen Anne's lace and bishop's weed, and mustard yellow goldenrod. And the fence that I was once so proud of has not weathered particularly well; one end has fallen from its post, and the whole enclosure looks rickety. There is a certain charm in that, I suppose.

As we walk along the stone wall and tree line, there's the old stump that we tried to start mushroom spore on, and there are the two vagrant apples that seeded themselves several years before we arrived, their branches reaching toward the light in the center of the field. There is a broad copse of young blooming wild black cherry, the flowers scented like crushed almonds and marzipan. Depending on the season, there might be the little white and lavender flowers of wild aster, or thick jungles of exotic-looking fern. Fallen logs here and there litter both sides of the wall, and halfway down there is a charred and hollowed-out trunk, a tree clearly once struck by lightning, succumbing to birds and insects over the years, a perfect spot for a wild bee colony.

The second half of the meadow, this expanse below the greenhouse, comprises still-virgin soil defined by buttercup in the spring, then later, red clover, black-eyed Susan, little clumps of sweet daisy, fleabane, threads of August-blooming goldenrod, and masses of wild mint let loose from the small clump my mother once gave me for our flower garden. This part of the farm, left mostly to its own devices, has a raw, yet soft beauty. In the Julys and Augusts of years past, the meadow used to host migrating monarchs. Thousands of orange-and-black tiger-striped butterflies would flock to the wildflowers, cling to the blooming mint, feeding like the bees on all the wild and profuse nectar. Butterfly enthusiasts used to come up our hill during the migration and make notes in old-fashioned science notebooks about the number of monarchs, the plants on which they were feeding, climate conditions, and other variables that might shed more light on the unique habitat and life of the monarch. We used to delight in the swarms of butterflies in the air and on the plants and how they brought an otherworldly and decidedly wild element to the landscape.

But that was over ten years ago. Since that time monarchs have been steadily on the decline. Each year we have seen fewer and fewer monarchs, until this year, when we saw maybe two the whole season. Even just a couple of years ago we felt encouraged, as we have a fair amount of milkweed growing in our block of Riesling. We found a handful of monarch chrysalises in all their black, white, yellow, and chartreuse striped glory reposing in the milkweed. We took photographs, we celebrated, we watched them flit as adults in and out of the vines. This year there were none.

The reasons why seem to be an old story now, and a repetitive one as well. The same kinds of farming methods that have so gravely affected the honeybee also seem to have wreaked havoc with the monarchs. It works something like this: Herbicides like Roundup, which are created to strip farmland of the non-monetized flora that compete with intensive plantings of a single crop, do their job perfectly and eradicate "weeds" like milkweed, which grows readily in the Midwest where there are many large farms. Monarchs have a unique and significant symbiosis with the milkweed, and it is the only plant on which they will lay eggs and larvae will feed. This evolution has occurred over millions of years. But if the herbicides eradicate the milkweed, where can the monarchs procreate? Apparently nowhere. Fewer milkweeds equals fewer monarchs.

GMOs also extend their hands into the debate because the monoculture crops grown for money have been modified not only with the Bt organism to combat pests but also so as not to be adversely affected by the Roundup

herbicide. Monoculture crops are engineered to be überplants. Since the herbicide can't kill them, the dosages for the herbicide are higher in order to rid the land of all the other plants that are feared as competitors.

I've heard many of the arguments in defense of GMO crops. One of the primary ones is that our planet and nature are amazing, and they will right themselves in the long run: They will adapt, the honeybee will adapt, the monarch will adapt. Yes, I too believe that our planet will adapt over time; we have witnessed its resilience in so many ways. But this kind of thinking strikes me as shortsighted at best. Millennia of adaptation occurred before the beneficial relationship fully developed between monarchs and milkweed. As a species, these butterflies cannot adapt quickly enough to another plant to prevent their own eradication.

Many will say that evolution itself is merciless, and that we as humans are part of evolution, and perhaps all this business with chemical farming and GMOs is just our particular brand of participating in evolution and adaptation. A good orator can make this sound quite plausible, but there seems to be an awful lot of evolution going on all at once, and aside from a quick and random natural disaster, experience doesn't suggest that evolution works in quite this accelerated way.

While the monarchs may never adapt to the disappearance of milkweed in time for them to survive, we watch our field adapt to the shifts in weather and the shifts in its own landscape. Even though this is a relatively wild corner of our land, everything we do in cultivating around it affects it. We are sensitive to the need for the wild among the tamed. We do have plans for this field: apple and pear trees, rows of vegetables, several stands of cosmos and sunflowers, maybe even some pigs and chickens to roam. But we know that there is a fine line between husbandry and nature, and that the two must coexist together for both to work happily side by side and for naturalistic farm to be fully bountiful.

The wild aspect of a farm is often overlooked, or mentioned only in passing, or described as part of another piece of the farm. But the wild is crucial to the cultivated. The introduction of things like cover crops or planting flowers and bushes that attract beneficial insects and birds all fall under the concept of cultivation. In essence these growing methods have been designed to mimic nature and her patterns. When we cultivate, we disturb an already established wild ecosystem. As stewards of our land it is our job to re-create or encourage the re-creation of that ecosystem in tandem with our cultivation. These are good and responsible ideas and ones that we follow, but we are also interested in learning how the natural wildness of our land coexists with our cultivation and can help us be better farmers.

Wild Plants on the Farm

There is a long history of using wild plants in agriculture. We can turn once again to documents like Virgil's *Georgics* for a glimpse into farming over two thousand years ago. Such knowledge, like the medicinal properties of plants, was long the norm in civilized societies. And while the Industrial Revolution has brought us many advantages in terms of how we live, there are also so many things we have lost—and most particularly, our connection with our landscape and how the wild world works.

WE ARE in an auspicious period now. The swing back to an intense interest in agriculture, to knowing where our food comes from and how it is grown, is both heartening and necessary. The back-to-the-land movement that began in the 1960s has a broader reach now; a whole new and younger generation has been inspired to take up the mantle. And while many food movements came out of the explorations of the 1960s, a larger group of young farmers is also concerned with the intersection of how the food grown responsibly gets to the plate. Cook-farmers and fermenter-farmers are becoming more prevalent in this pie wedge of our population. Things are definitely looking up.

Along with these cross-interests there is an abiding respect for the nature on the farm and how it is part of the holistic farm agriculture. The use of wild plants in agriculture is starting to gain a foothold with increasing interest in biodynamic farming. In many ways I think we have biodynamic viticulture to thank for this. The use of wild plant medicines and plant boosters in the vineyard has long been part of natural, organic, and biodynamic vineyard practices in Europe. Now, we see them in the new waves of winegrowers on both the West and East Coasts of our own country.

Biodynamic farming is centered on both plant- and manure-based preparations and treatments. There are the core seven wild plants used for making teas and decoctions: horsetail, nettle, oakbark, valerian, dandelion, chamomile, and yarrow. All are plants readily found in the wilds of Europe where the codification of biodynamics began. In our climate, too, all of these plants are found in the wild, except for valerian. However, valerian is cultivated in both Vermont and New Hampshire in quantity for medicinal purposes.

Walking down the dirt road from the barn to the entrance end of the vineyard, the hedgerow offers an abundance of horsetail, nettle, and white yarrow. We have a few oak trees on our property, but down the road is a proliferation of oaks of all different sizes. Dandelion is everywhere. Native wild chamomile grows in our drive and peastone garden paths—it likes stony ground. It is not the same plant as the chamomile found in Europe or the variety cultivated typically for tea, but it makes a beautiful pineapple-scented tea, and it is native to our soil.

Because we can cultivate valerian easily here, it is planted near the greenhouse. It is easy for us to make teas from these seven plants to use in the gardens and vineyard. Certainly chamomile, yarrow, valerian, and dandelion can be cultivated, and these plants do grow here and there throughout our gardens, but they are all so readily available in the wild that growing them intentionally isn't necessary.

Each plant has its purpose on the farm. The roles have been somewhat standardized now for their use: Horsetail is good for managing mildews, and nettle is good for sap circulation. There are relatively specific recipes for how to make teas, and many years of research by others far and wide has been done on the efficacy of these plants on a farm.

The Role of the Wild on Farm Health

The wild plants that we turn to are often considered plant medicine and used as such to help balance the equilibrium on the farm naturally. Some growers who have pursued a biodynamic growing program sometimes come to the conclusion that implementing all the natural and wild teas and sprays is not necessary; if the vineyard or farm isn't actually sick, why give it medicine? A valid point, but I also believe that the tea sprays can be used as boosters, not just medicines, and that such decisions are completely dependent on the year and the state of balance on the farm. When biodynamics became structured, farmers and philosophers saw that farmland had already begun to suffer from industrial and chemical methods and thinking. Many of these plant medicines that are at the core of biodynamics had been in use long before the Industrial Revolution. We, as farmers, had lost touch with the old art of agriculture. Biodynamics

set out to systematize those methods and to expand upon them.
For biodynamic practitioners, all lands and soils were "sick" or
out of balance. A lot of land needed to be converted from conventional farming back to more natural methods. This takes time,
and there is a procedure to follow. I don't think anyone who has
gone down the road hand in hand with biodynamics, even if they
no longer actively implement it, would disagree that the procedure is very effective. But then the question becomes what to do,
or what to implement, once your land becomes a more naturally
and cleanly functioning organism? Do you still need to treat it in
the same way?

SOME CLIMATES AND LANDSCAPES ARE relatively pathogen-free and can
enjoy a more hands-off approach to their organic farming. Because of sun,
lack of humidity, length of season, and the rockiness of the soil, sprays and
cultivation may only be minimally used. (I am thinking of southern France
as I write this.) But other climates will always have difficulties unless they
have an exceedingly propitious year weatherwise or they are situated in a
pocket of the landscape that is largely farmed naturally. If your farm is next
to a conventional farm, you will always need to confront the drift of chemicals onto your land.

In our corner of Vermont, we experience great heat and full sun and our
growing season seems to lengthen every year, but we do have high humidity,
and we have had a string of very rainy early summers—perfect conditions for
many diseases and predatory insects. Our land has never been chemically
farmed, yet we grapple with any number of nuisances each season. And even
though our specific land has not seen industrial-style agriculture and we live
in a very rural state and our village supports a very active organic agriculture

movement, the effects of what happens elsewhere do impact us. We see it in the changes in the wild landscape of flora and fauna that surrounds us. The decline of the monarch butterflies is a perfect example.

Having farmed with biodynamic and organic methods for a few years now, enough to see the effects, I believe the use of the wild in cultivation acts as a sort of inspiration to tended plants and land, that it reminds them in ways we cannot comprehend what it once was to be wild, to survive in a world with no human intervention on their behalf.

Many growers and laymen see biodynamics as a New Age philosophical quest and liken it to modern witchery and voodoo. Because Rudolf Steiner spearheaded the collection and systemization of antique agricultural methods into biodynamics and he was also a spiritual philosopher, an automatic assumption is that it is only for the spiritually minded. Certainly, many gardeners and farmers who do approach their lives on a spiritual level find biodynamics attractive. Biodynamics can be mystical in that it discusses and addresses the bigger picture of our life here on planet Earth and how growing our food is at the very core of our existence. Without food, we do not survive. But biodynamics also looks at the details, the microscopic life of the soils, plants, and animals both in the wild and in cultivation. When you grow a plant from seed or witness the birth of an animal, whether you are spiritual or not, I do not think you can deny that there are other forces at work than our own human efforts. The life cycles of the natural world will proceed very well without us.

Biodynamics comes down to the life forces of what is natural and what is wild. The methodology actually examines in great detail the physical working of plants, soils, and atmospheres. Organic thinking, of which codified biodynamics is an outcropping, examines the plants and the soils as well, but I think biodynamics becomes revolutionary when it examines how plants and animals and soils respond to the workings of physics and microbiology: the cycles of the sun and moon, the cellular architecture of plants and animals, the effects of minerals on and in flora and fauna.

When I began studying biodynamic viticulture, I was stunned by how the precepts of physical science supported the patterns of preparations and applications. A whole book could be written on the appearance and uses in the plant and animal world of the mineral silica alone. Silica occurs in plants, and in our mammalian bodies; it occurs as a mineral element in the crust of the Earth. We have eyes and are able to see because of silica—it is literally the lens with which we as humans can view the world. It is the primary component of the cell—and of smartphones, which have become so integral to our lives. The silica in the phone captures and emits the microwaves from cell towers, just as the silica found in the wild, prehistoric fern horsetail captures the heat and light of the sun and transmits it to the plants it is treating as a drying force, a ripening force, when it is made into a plant tea and applied. In this way, biodynamic agriculture harnesses what is wild and uses that to inform and transform.

The Botany of the Wild

The wild world around us can tell us so much if we just pay attention. In the chapter on the vineyard, I wrote about my experience in France studying biodynamic viticulture and how the existence and botany of wild plants within the vineyard setting can and should inform a winegrower's plan for the season. It is the same in any component of the farm. What grows wild in the orchard, or the vegetable garden, tells us about the nature of the soil and the ecology of the landscape. Modern organic methods rely on the idea of forest-edge ecology to inform the design and practices for a specific landscape. The concept looks to the wild to create a cultivated world that operates on wild precepts.

NOT ONLY CAN THE wild alpine strawberry growing at the top of the meadow tell us that our soil is very acidic there, but the natural presence of red and white clover in the vineyard suggests that the soil is rich in nitrogen and will become more so as we mow and work that clover back onto the ground. We have never seeded clover there. The milkweed likes sandier, stonier, well-drained soil and only appears in our block of the vineyard where we have Riesling planted and the land slopes in two directions. This is one of the reasons that we work with cover crops rotationally. Every three years we don't plant cover crops, to allow the soil to produce plants that find it hospitable. Some of these plants will come up from seeds from last year's covers, but the mix of plants will tell us what the soil is doing and what kinds of cover crops we need to plant the following year, or where we need to add compost or work the soil more thoroughly. The season's succession of plants will also tell us what nutrients are present in the soil throughout the season.

The personalities and botanies of these wild plants bring something to the land as well. That presence of dandelions attracts sunlight with their yellow, sunny faces, and their roots till the ground naturally, breaking up the soil and creating space after they die back and rot, allowing both the micro- and macroscopic life in the soil to travel more freely. Certain plants attract beneficial insects. That wild blackberry that runs along the top end of our brook brings the Anagrus wasp, which is always hungry for the grape leaf-hopper as well as predatory mites that can destroy the vineyard. The study of these plants with a biodynamic eye can also provide interesting adaptations of plant medicines. The natural world works in symbiotic and mysterious ways. Comfrey always grows nearby stinging nettle; they are natural companions. And comfrey provides the antidote to the sting of the nettle. Just rub the affected skin with comfrey leaf for relief.

Ultimately, the soil knows better than we do, and what the environment needs it will produce. Again, consider dandelions: They appear in their carpet of golden spring glory when you have clay-heavy and compacted soils that need breaking up. Nature provides its own clever tillage. Nature is always striving for balance, always adding and subtracting to keep the scale in equilibrium. And nature does not discriminate; it takes us and all our additions and subtractions into account. It responds when we build a structure in its midst; it responds to the crop we grow in the wild meadow. It calibrates those responses with these new additions in mind. There is an uncanny consciousness happening in ways that we can see, but also on a microscopic level that we cannot see. Nature struggles and can retaliate when the human hand tries to eradicate nature's efforts. When the industrial, large-scale farmer tries to destroy the corn borer, the corn borer very quickly adapts to all the pesticide tricks and becomes immune (unlike our poor monarch, who wants nothing whatsoever to do with corn and is slower to adapt). In this mind-set, the farmer must find another, more lethal way of getting rid of this pest, or learn to alternate pesticides in the hope that the corn borer won't understand the pesticides too well and learn to circumvent them. This kind of farming becomes an escalating battle between man and nature. Haven't we learned yet that we will never win? In the end, I believe it is we who will suffer the ultimate consequences of consistently aggressive and fear-based agriculture.

Edible and Useful Wild Plants

As we walk to the spot lower in the meadow where our oldest apple tree stands, I freely marvel again at nature's way. The old apple, in need of more pruning in order to help it balance itself, makes its own efforts. The healthy side of the tree grows an abundant amount of shoots reaching for the sky, a mass of green leaves calling down the sun into this little fold in the field. We can see from its form and growth patterns how we can help. By removing a few scraggly trees that are growing too close, struggling toward their own demise, but that still block the light, we get rid of their deadweight and help the healthy tree find room for more relaxed growth. We see the quest for survival on this forest edge: the larger, older apple tree grappling with these younger, upstart trees, who sense weakness and compete for dominance.

THIS APPLE TREE, along with others that have cropped up along the old farmer's hard-won stone walls, yields a wild harvest. The bitter, sharp, and tannic fruits provide another kind of balance: They broaden the flavors of our farmhouse cider. Wild fruits and plants not only contribute to our farm by educating us as farmers or by contributing to its cultivated companions, but also feed us through our menu here at the farm and at the osteria. The wild dandelions give us tasty and bitter greens for salads, braises, and soups. The fiddlehead ferns, a Vermont delicacy that grows at the bend in the brook,

WILD LEEKS (ALLIUM TRICOCCUM)

WASHING WILD LEEKS

PURSLANE

MUSSELS ROASTING UNDER PINE BOUGHS

provide one of the first of our spring dishes. Well blanched then sautéed in white wine and olive oil with a little garlic, they make a beautiful salad with some shaved hard cheese and our wild and spicy arugula. The wild leeks, or ramps, growing on the wide bank down by the big stone, which looks like a giant's plaything, give delicious rise to other spring dishes: soups, pastas, or simply roasted and served with just a swirl of olive oil and a squeeze of lemon.

Up from the brook, just on the edge of the vineyard where the birch and poplar branches hang over the field, honey mushrooms proliferate. If we catch them before the bugs do, these flowers of the soil are sautéed and served on toasts or polenta or with fresh, silk noodles we roll out or in a risotto or farotto, a dish made with the nutty kernels of farro, one of the oldest forms of wheat.

In August the blackberries and raspberries begin to ripen along the brook bank in the sun, and at the top of the meadow, again under the trees, the honey mushrooms will proliferate among the lavender and gold alpine and old field asters. Here, also, we still find those wild alpine strawberries that are so sweet and small that there are never enough to make anything except a small addition to a dessert, providing an unexpected sweet burst of flavor. The native thimbleberry or purple-flowering raspberry grows near the brook too, and its flat raspberry structure makes a delicious snack on a hot July afternoon in the vineyard while we rest in the shade near the water.

Wild chicories and chickweeds wend through the vineyard, providing tender leaves for salads. In the vegetable gardens proper, succulent purslane spreads out on the bare soil in curls and flourishes, also providing a crunchy texture to those summer salads. Nettle, queen of plant medicines and boosters on the farm, also is delicious in pastas and soups, or braised beneath roasted sausages. The pines that have grown up on the edge of the field in certain spots make for great infusions in vinegar or brandy or honey, or for roasting meat or seafood in an open outdoor fire. The needles also make for a tonifying tea, as does the native chamomile and the statuesque mullein. The wild rose hips also provide for tea as well as for jelly or for roasting. And the wild rose petals go into syrups for cooking meats or desserts or for infusing brandy.

The bounty of wildness at the edges of our land is rather staggering, once we reflect on how much it feeds us and the farm.

The Hop Hornbeam Wood

At the very bottom of our pie-shaped piece of land is a secret garden. A garden of trees. Nestled into a corner of stone wall between the brook and the forest is a little grove of hop hornbeam. The trunks are tall and narrow and sinuous. The bark is pale gray and finely woven with silver scales. The root structures look like dragon's feet. *Ostrya virginiana* is in the birch family and is often called ironwood, as it is incredibly tough and durable and hard to saw through. The leaves are oblong and double-toothed like birch leaves, and the catkins form clusters reminiscent of hops for beer.

OSTRYA COMES FROM the Greek word *ostrua*, or "bonelike," referring to the hardness of the tree's wood. The fruit from the catkins are used as food by the Lepidoptera species like the winter moth, walnut sphinx, and coleophora. Many foresters think of this tree as a weed and as useless, but like most weeds, it has its place in the naturally evolving ecosystem of the forest.

For us on the farm, the hop hornbeam wood is the transitional place between farm and forest. We gather mushroom for the osteria and farm dinners in the larger forest, studying and learning more about the wild environment that surrounds us. While we haven't seen many mushrooms in the hop hornbeam wood, there have been a significant number of fallen trees due to storms and overcrowding. In our efforts to manage that part of our land, we have pulled out fallen logs and used them for fences and trellising material. Given their hardness, it seems they will last forever.

OVOLI MUSHROOMS IN THE FOREST

On our petite alpine farm, we stand at this forest's edge named the Chateauguay. Our forest is teeming with plant and animal life: an ancient maple grove sits upon the hill above us, shrouded by hemlock, spruce, and pine. Birch, copper beech, ash, poplar, wild black cherry, and wild apple define the deciduous trees, underpinned by fern and moss, trout lily and trillium. Oyster, lobster, ovoli, scented coral, and white comb's tooth mushrooms bloom after the right combination of rains and heat. Brindled coyote, black bear, white-tailed deer, moose, fisher cat, porcupine, raccoon, possum, white ermine, fox, black-as-night crows, hawks and falcons, wild turkeys, snowy owls, and countless other small birds all cross paths and live along established routes and habitats. We are but a small cultivated opening in their otherwise wild and obscured landscape, and we must cross their paths and lives too. We share in this forest and meadow, and we must understand something of one another if we are all to get along. We must be prepared and know that what

OVOLI MUSHROOMS IN THE FOREST   WILD TURKEY EGGS IN THE HEDGEROW

we cultivate may some-times be of interest to our wild companions. The bear will eat the honey, the deer will dine on apples, the birds will feast on grapes, the fisher will massacre the chickens, the  raccoon will steal from the garden or the soup pot. I have stood in my vineyard and cried after the sparrows descended on the fruit and picked the bunches almost clean in twenty-four hours. But I cannot blame the birds. We were the ones who didn't get the bird netting on in time. We didn't set the boundaries. The birds are only doing what comes naturally.

ALL WE CAN DO in living with so much wild around us is provide protection for our gardens, vineyard, orchard, and bees and try to be conscious of not encroaching on the wild fauna's wild food. Just as we leave enough honey for the bees in winter, we must share in the wild.

And in the wild we delight.

Cantina

At the end of the growing season comes harvest. Some harvests are ongoing; others happen once. Our harvests have two hearts: the kitchen and the cantina. Fruits and vegetables go into the kitchen. Grapes, apples, and we hope one day pears come to the cantina.

THE WORD *cantina* comes from the Italian. It means "cellar, winery, cave." Here in the United States we know it also to mean a type of bar in Mexican, Spanish, and Italian culture. A place where people (in Mexico, often only men) go for food and drink. We like the double meaning, and our adoption of the word came swiftly and without much forethought. The space where we were fermenting and storing the wine became known simply as the cantina; I'm sure it's a holdover from how we learned to talk about wine cellars when we lived in Italy.

Even though the harvest signals the end of the obvious growing season, it is not really the end of the growing season. Work in the cantina is an exten-

sion of everything that has happened in the field. The season will express itself as the grapes turn into wine and the apples into cider. Our job in the cantina is to provide a clean space for the fruit to ferment into wine.

Here, at this point in a grape's life, is where philosophies and ideologies about fermentation and the making of wine begin to diverge. In current cellar practice and the writing about cellar practice, there are for the most part two camps firmly situated on opposite banks. There are those who believe that the point of growing wine, of fermenting grapes into wine, is not only about having something enjoyable to drink with our food, but also to express something of a place, that sometimes illusive terroir. This belief equates hands-on, naturalistic farming to art and craft. They believe that wine becomes a looking glass into the year of a place and its people. Wine has a story to tell about a specific point in time. This camp often gets labeled under the rubric of *natural wine*. Even the name itself can elicit strong reactions from those both within and without this group. Some feel the moniker *natural wine* is not explicit enough. What does it mean? How do you codify it? Natural can mean very little on American shores where marketing teams have co-opted it to convince the public that they are buying healthy, pure, responsibly made products. In the United States, natural can actually mean none of this. All-natural chicken can be made from chickens who are raised in cages, whose feet have never touched the ground, and who are fed a heavy diet of corn. These chickens may have no artificial ingredients and no added coloring and they will have been minimally processed—all steps in the right direction—but this is where the USDA definition stops.

Fairly recently, some wine appreciators have started to turn their attention to where their wine comes from and how it is farmed, but there is still a large portion of the wine-drinking population who believe that wine, in and of itself, is naturally a natural product. But as more and more people experience wine cultures abroad, and drink local wine freely with their meals and do not suffer headaches or feel hungover the day after, questions about the use of additional sulfites in wine and the use of chemicals in the farming as well as in the cellar have arisen. Wines that are consumed in the native wine culture and that are not exported tend to have few to no sulfites used at bottling because the wines won't travel more than a few miles. Concerns about preservation and aging are less, as these kinds of wines tend to be drunk early on in their lives. Even if the use of sulfites, or other additives that may actually be the culprits in the wine, doesn't cause the headache directly, when a traveler comes back to the United States and resumes his or her relationship with a mass-marketed wine that is brought home from the grocery store, the traveler feels and tastes the difference. It is one of the most frequently asked questions at our osteria: "Why do I feel great when I drink wine in Italy, France, or Germany, and when I come home, wine makes me feel bad?" My answer is always that wine in the United States doesn't have to make you feel sick. You can have the same experience here that

you have had abroad with your wine. But it requires a little extra work on your part to learn about the producers, their wines, and their farming and cellar habits. You research your food—where it is coming from or how it is grown—and researching your wine is no different. Wine is just another form of food, and it should be considered as such. And winegrowers should be held accountable for the way it is grown and made.

In wine, both those for and those against the natural wine concept have concerns about how to police whether a wine producer is really operating under "natural" rules. Anybody can say "we work naturally," "we work sustainably," "we work organically" and not really mean it. And the consumer, by and large, wouldn't really know the truth. There are those in the wine world that call for the codification of "natural," a certification process so that the average wine buyer knows exactly what he is getting into.

But certification is tricky. We have biodynamic certifications, organic certifications, sustainable certifications, and conventional-converting-to-organic-middle-ground certifications. Some proponents of certification support the process, as they believe that we need to have rules in place to be followed to ensure that the farming is what it is said to be. Others feel certification hamstrings the farmer's choices, requiring the farmer to follow a specific code that may or may not be in the land's best interest at any given time. Meaning that, for instance, if you have something like a compost amendment requirement for each year, maybe your soil doesn't need compost, and it will in fact imbalance your soil to add it; but if you don't make the addition, you will lose your certification. And yet another group of people believes that the standards aren't strict enough. All this discussion is well and fine, but the reality is that certification is expensive. Many small farmers, the ones who really believe in making wine as cleanly and naturally as possible, and the ones who actually do the work, can't afford the ticket price of such an outlay to be able to print "organic" or another like-minded term on their label.

Organic certification in the Unites States actually is fairly stringent and extends to the cellar as well as the field. But label laws have allowed producers to skirt the cellar requirements by labeling their wine "organically grown" or "made with organically grown grapes." These laws require that only 70 percent of the fruit be grown organically, and there is no provision for the work in the cellar.

When you are purchasing something like wine, and you look for certification, you will know that the farming follows the certification requirements. But the certification is generally only for the agriculture in the field. It does not necessarily address anything that happens after harvest. You could have the healthiest of vineyards, and still use whatever number of additives you like once the fruit comes in out of the field. An organically certified wine is no guarantee that the producer is organic in the cantina. All that hard work spent focusing on the balance of the vineyard and the vines' ability to reflect the terroir can go right down the drain.

Most people who appreciate wine but who haven't had a lot of experience in understanding the growing and making of wine are surprised to hear that what can happen in the winery isn't always a natural process. According to USDA standards, I've counted over seventy additives that can be present

BOTTLES ON A FRENCH BOTTLE DRYER

in wine: calcium alginate, silicon dioxide, edible gelatin, gum arabic, poly-vinul-polypyr-rolindone, activated charcoal, copper sulfate, tannin, saccha-rose, calcium carbonate, thiamine hydrochloride, sorbic acid, ascorbic acid, dimethyl dicarbonate, and water, just to name a few. These additives are not necessarily bad for you, and many of them come from natural processes. And wines made with these things don't necessarily taste bad; in fact they may taste "good," though this is such a subjective view (and you may have a heck of a headache after drinking a bottle). The point here is that a wine treated with these additives is no longer the wine that was made solely and natu-rally from the grapes grown in the field. It is no longer a representation of a place or of what the farmer grew. Purity of ingredient no longer matters. It is rather a manipulation of chemistry in the laboratory.

There are some wine critics and winemakers who believe terroir doesn't exist, and that to worry about the expression of a particular season and a specific piece of land is absurd. These same critics and producers also believe in the efficacy of the lab. Wine is "corrected" based on what the laboratory numbers provide. If the acidity numbers seem higher than is desired, the producer can deacidify in the lab. If the sugar is too high, and therefore the ultimate alcohol is too high, a cellar master can add water to dilute the wine. Sulfite can be used from beginning to end, to kill native yeasts, to correct scent and taste problems, to correct the proliferation of acid bacteria, and to preserve the wine in the bottle. But when sulfite is used to this extent, it strips the wine of character, so character must be added back in as color, aroma, flavor. It is wine made by numbers, which seems fail-safe to some and is the smartest economic model. There are even some producers who contract for organically grown fruit, or grow their own, and then rely on their lab to make the wine. They've fulfilled their duty as a grower and a marketer by using organically grown grapes (and, again, 30 percent could be conventionally grown for high yields), and they don't have to worry about any difficulties in the season or about lesser fruit; they can tailor-make the wine once they get into the winery. There is a correction for everything.

This scenario begs the question: Is this real wine? If we apply the same idea of painting to paint-by-numbers kits, we can see a moderately accomplished finished painting by someone who doesn't really have the art or the craft to paint. We don't think of people who paint by numbers as painters, and we would never categorize their finished work as a real, original painting. Wine made by numbers may be acceptable to the taste, may even be enjoyable in the moment, just like the painting may be pretty to look at, but why would we consider it to be real wine?

Real wine seems to be a more applicable name these days than *natural wine,* and natural wine ambassadors like wine writer Alice Feiring use the term *real wine* when talking about wines that are made as naturally as possible and with as little intervention as possible. Her thorough and engaging newsletter *The Feiring Line* carries the subtitle *The Real Wine Newsletter.* It is most important to note that human intervention is necessary to grow

PIGÉAGE OF GRAPES

wine, or any other agricultural endeavor. But supporters and practitioners of real wines make a distinction between guiding the process minimally and overt manipulation that ultimately changes or homogenizes the character of

RACKING WINE IN THE CANTINA

the wine, minimal intervention is about letting the wine do as much of its own work as possible and tell its own story.

I like the use of both *natural* and *real*, and what I like about both terms is that they are deliberately specific and broad. Natural wine was originally coined by the French as *vin naturel*, and it was meant to indicate wine that had been made with as little intervention as possible both in the field and in the cellar, with native yeasts and little to no sulfites. Alice Feiring's book *Naked Wine* examines this issue carefully and explores the instigators and evolution of wine made in this fashion, which in reality represents a return to the way wine was made before the Industrial Revolution. I like the idea of natural wine because it doesn't pin down the definition. It allows for a variety of approaches and philosophies. Wine grown organically, biodynamically, even with *lutte raisonée* can all fall under the big-tent definition of natural wine. Lutte raisonée, or the "reasoned battle," means that a grower tries to work without chemicals and pesticides as much as possible but will use a chemical tool to combat a specific problem that puts the vineyard or crop as a whole at risk and then will return to natural procedures. Instead, I wish there was another term for lutte raisonée. The idea of "battle" when talking about real wines seems the wrong sentiment. Wines made with no additional sulfites and wines made with a small amount of sulfites at bottling are both allowed in the natural and real wine discussion.

I like the term *natural wine* because it promotes the shared belief that wine is grown in a vineyard free from chemicals and pesticides and that the hand of the producer intervenes as little as possible in the cantina. It promotes wild, indigenous yeasts and little use of sulfites. At the same time, it allows every producer to respond to his or her season and location without having to follow the same set of prescriptive rules. A producer can use a horse or a tractor to plow, sheep can graze in the vineyard, or the rows can be mowed. A grower can use nettle, copper, sulfur, baking soda, milk, or horsetail in a spray application. Compost can be applied when needed. The producer can till or not till. In the cellar, the winegrower might employ open-vat fermentation, or fermentation in used barrels or in clay vessels. There might be *elévage* in glass demijohns or concrete eggs. The wine might be made using carbonic maceration (the exclusion of oxygen in the process of whole-berry, cluster fermentation, meaning that the fermentation begins within the skin walls of the grape before expanding outward) or in open containers that rely on a layer of carbon dioxide created from the fermentation, or a naturally occurring *flor* or bacteria to protect the wine. While many producers who fall in the natural wine category prefer not to use stainless steel or fiberglass fermenters, they at least have that option if needed. Sometimes these are the most economically viable choices for a small producer.

Using the term *real wine* fits the same model of a general belief and principles that encompass many different paradigms to achieve like-minded goals. When someone brings to me a natural or real wine, I have a general idea of what to expect: a wine that was farmed in some way with organic intentions, fermented on native yeast, with little to no filtering and few to no sulfites added at bottling. Part of the fun of getting to know the wine is appreciating and learning the craft and skill the producer employs when out in the vineyard and how she approaches fermentation in the cellar.

Many wine writers, critics, and producers call for a more specific delineation, if not certification, of *natural* and *real*. There are those who feel the word *natural* is too loose and doesn't really inform you about the true nature of the wines. But the reality is that none of the words do. If you really want to know where your wine comes from, you must do the work to trace your wine back to its source, just as increasing numbers of us now demand traceability in the food we eat.

I understand the desire to define real or natural wine precisely, but as someone who grows wine and educates about wine, I find the effort too restrictive. Growing and fermenting wine naturally is a fluid process and the words used to describe it should also be fluid. This allows for the sharing of ideas and practices and leaves room for different ideas and practices, as well as the evolution of those ideas and practices. We already have plenty of certifications for particular kinds of agriculture, and we need to employ a broader term such as *natural wine* or *real wine* to aggregate all these specific methods for getting at the same goal.

When I first began making wine, I never imagined that I would be able to enter into the natural wine dialogue as a producer. The impetus for going through the physical actions of making wine was for my own education and edification. I wanted to know what it felt like to make wine, even in a tiny batch. I wanted to watch the fermentation, stir the lees, taste the development of fruit into wine. One of the things I really wanted to understand was how to better taste unfinished wines and gain more understanding in how they might develop as finished wines. In visiting other producers who made the wines on the osteria's wine list, not only did we taste bottled wines in the tasting room or around a rustic table with food, but we would taste straight from the tank or barrel. Young, still fermenting wines taste very different from wines deemed ready, and I wanted to understand how to gauge how a wine might develop, what kind of indications the young wine might give to imagine the finished wine.

IN ORDER TO CONDUCT THIS EXPERIMENT, my process was not particularly natural. I bought fruit or juice from a produce purveyor, and I began fermentations in 5-gallon buckets in our clawfooted bathtub. It was easy to keep things clean. Because I didn't know anything about the vineyard that had grown the grapes or their practices, I used relatively neutral cultured yeast. (When you don't know how healthy the native yeasts might be from the vineyard, wild fermentations can fail or turn the wine to vinegar.) By my third year of making wine I was able to buy a few baskets of fruit from another winemaker in Vermont, so I could try my hand at producing something from local grapes and with native yeast. I was still working in the bathtub. By the fourth year, we had gotten serious. We knew we would have apples from our orchard to make cider, and we had found a grape grower willing to sell a little bit of fruit to us in addition to all the other producers he supplied. Through a good recommendation from another winemaker, we were able to get our foot in the door.

That fourth year, we knew I would have to move out of the bathtub. And because we knew we had access to local fruit, we wanted to be official. If the fruit turned into wine and not vinegar, we thought how wonderful it would be to have our own wine and cider at the osteria. And if it turned into something more appropriate to salad dressing, we thought how lovely it would be to cook with our own house-made vinegar.

Our spatial options were limited. We had the one barn with one small space at the back that was insulated enough to be a year-round cantina—a small space that was supposed to one day be Caleb's woodworking studio. We commandeered this space for the winery. We bonded and licensed the shed space for wine and the barn itself as a seasonal tasting space.

We've worked with what we have. Our friend who makes wine just down the hill from us has a great old stone-lined cellar beneath his house in which he makes his wine and stores it. Cellars are best for wine, especially in our climate, as the temperature stays constant throughout the year. But our own cellar was too crowded with our unwanted and stored belongings, plus you have to go through our bedroom to get there. Somehow we didn't see this working in the long run.

So, we've adapted and created a space that works for our small size. The cantina reflects the word *microwinery* perfectly. We've produced in numbers of bottles rather than cases. And, even as we grow, we will always be a microwinery, probably never producing more than a thousand cases or so annually. This is fine with us. It is in keeping with the size of our farm, the nature of our Vermont landscape, the diversity of our work and crops.

Fermentation

I return to that notion that fermentation and fire are the two hallmarks of civilization. Both take food from its basic and animalistic elements to something more cogent; food goes from its raw state and is transformed into something else that requires contemplation and some skill. From an evolutionary level, cooking by either fermentation or fire means the food is easier for our human bodies to digest, and these processes provide easy access to the energy stored in the food. From an artistic level, they allow us to shape our food in more complex ways, providing intricate

PUNCHING DOWN FERMENTING GRAPES

ROSÉ JUICE FROM HARVEST

and intriguing flavors. Fire and fermentation also provide creature comforts, essentially that of heat. Fire gives off calories in heat, and fermented foods create inner heat. Both fire and fermentation improve our basic, animalistic lots in life.

THE FERMENTATION OF WINE is the complement to photosynthesis in the vines. Photosynthesis is the transformation of heat and light into matter. In the cellar, fermentation is the liberation of heat and gas. Inhalation and exhalation. Birth and death. Transformation of one substance into another. A life cycle.

This is one of the basic reasons that many wine appreciators, growers, producers, and writers believe that wine begins in the vineyard. It begins with the photosynthesis in the leaves and the formation of the flower that turns into the grape and the cluster. Fermentation becomes an extension of the vineyard work, and just as important. The yeast that grows on the fruit and exists in the cellar is part of the culture of the vineyard. It is dependent on what you use on your plants and fields during the season. Chemicals in the field—insecticides, pesticides, and fertilizers—all kill ambient yeasts as well as suppressing natural biological activity. But if you use minerals, plant teas, and compost, the yeasts of that season thrive. They come onto the fruit and into the cellar, and they ignite the fermentation process if they are allowed to do so. Every year they are slightly different, responding to the nature of that year's weather and growing conditions.

If you are a producer interested in preserving terroir in your wine, you are interested in preserving and encouraging your native yeast for fermentation. You consider the yeast created in the vineyard as part of what makes the wines unique. Even producers who use cultured yeast still hope to give something of the place the wine was grown, and will use what is considered a neutral yeast that doesn't necessarily impart specific qualities to a wine. And then there are those who don't believe in terroir, who promote the use of what has come to be known as "designer yeast." There are over two thousand strains of yeast that a winemaker can use to "create" a wine, giving it a certain flavor and texture profile. I live in Vermont and work with cold-hardy grapes, but if I wanted to make a wine from Vermont that resembled Cabernet Sauvignon from Napa Valley, I could engineer such a wine in my little cantina. I could choose a yeast for fermentation that would provide all the hallmarks of that particular kind of Napa Cabernet Sauvignon I was hoping to emulate, a varietal and climate so far removed from our alpine ecosystem. With such a yeast, along with some other cellar manipulations, the wine could turn out to be a very close approximation of a California Cab. It might even be very tasty. But it would say nothing about Vermont, or my farm, or my fruit, or me for that matter. It wouldn't say anything about California either. But to make wine like this is in no way my desire. Napa Valley Cabernet should be made in and come from Napa, and if we are going to call something a Vermont wine, that wine should be made here in Vermont with Vermont fruit. In a world where homogeneity is becoming more and more the norm, why would we want to make everything taste the same? Or from somewhere else? Why wouldn't we want foods and wines to be distinctive?

Why Landscape Matters

In the work that Caleb and I do at the osteria and the farm, the nature of an ingredient and where it comes from are important. Certainly, we can understand why the growing practices of a farm or a vineyard are critical in how they affect the landscape, the ecology, the environment, the plant's nutrient level, and the people. We understand that eating locally provides more for us and the environment than eating food that must be shipped thousands of miles. But this idea, that the specific location of where a food or wine is grown is important, can appear esoteric and distant from the more significant issues that face us in this world. Why does a discussion of terroir really matter? Why do we care? What difference does it make if all red wine tastes pretty much the same and all white wine follows suit?

NOTIONS LIKE TERROIR and landscape matter like art matters. These notions express themselves through food and wine in ways that affect us humans at our core. When we taste them, we understand something has shifted, something has changed, something is different. That may be all we understand, but we know we are moved. This kind of experience connects us more firmly to one another, as well as to the place in which we live, or where we visit, or that we imagine.

Terroir matters like history matters. It is part of history. France, Italy, Germany all have long histories of wine being made in certain places in certain vineyards. Some vineyards are deemed to have great terroir, others

less so, but they all exhibit some kind of local character when grown in a way that highlights the expression of that character. Government organizations like the AOC in France and the DOC in Italy, while flawed by politics, were created in order to protect the unique qualities that have developed over centuries in wines from certain places. They are ruled by a set of standards put in place largely to protect the patrimony, the history of these places in which one facet is defined by wine.

Since the founding of the international Slow Food organization in the mid-1980s, a whole network of protections has developed surrounding certain foods and raw materials, not just wine. The Slow Food Foundation for Biodiversity, which organizes and oversees these protected traditional products (called Presidia, after the Latin word meaning "fortress") is not unlike the UNESCO World Heritage organization, which protects and preserves certain places and structures that are of historic impact. In this time of GMO seeds and the intense hybridization of animal breeding programs, the Slow Food Presidia stand in place to preserve heirloom seeds and races of animals that ultimately form the foundations of a culture's cuisine. And these Presidia are not only confined to western Europe. They occur all over the world, preserving typical ingredients and heirloom dishes that define something about the life and the people of a place.

Eating a centuries-old recipe made from ingredients grown in its place of origin, the ingredients of the same quality and genetic makeup that they were one hundred, two hundred, three hundred, or more years ago, or drinking a wine that is made the same way and from the same place it was produced centuries ago, can be a very powerful and meaningful act. In these instances we are eating and drinking history. It is the one and only experience that can link us completely to our forebears. It provides the closest and most intimate understanding we can have of history—sensually, not just intellectually.

Terroir matters like stories matter. The words *history* and *story* are interlinked, and in some definitions and languages there is no difference. In English, the origin of the word *history* is Greek, connecting to the root word *histor*, meaning "a learned or wise man." *Historia* in Greek became the word for "narrative," for "finding out," and then for history as we know it. In Italian, a fictional story and history are defined by the same word: *storia*. In French, both are subsumed by *histoire*. Food, wine, and landscapes tell the stories of places and the people that live in them, the narrative encompassing past, present, and future. And like all stories, they educate, they motivate, they record and preserve.

These kinds of experiences and concerns are what make our human culture beautiful and complicated. Our modern culture, just like our farming techniques, should be diverse and maintain aspects of history and place while also improving on what we have learned. If we delete our history, if we delete terroir, if we let parts of our heritage become extinct, we learn nothing about how to be now and how to move forward.

So, we grow food and wine on our small 8-acre farm in order to preserve something of our human history and how it connects us to the people we have lived with and the places we have experienced and lived in, to which and to whom we are attached. We preserve to revitalize and inspire ourselves and those around us. We look back to the farming of our forebears to inform how we create our own unique and whole farm agriculture that exists in this place at this time. We look to the wilds around us to also influence the choices we

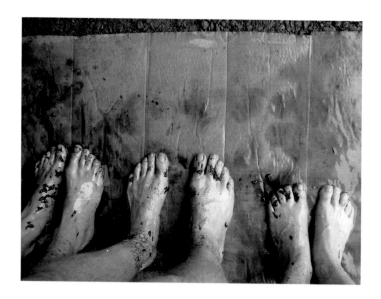

make in our growing season. We look at farming like we look at our kitchen and cantina. We turn to old cookbooks and the oral recipes of the elderly that have been handed down over generations as the bulwark of our menus at our home farm and at the osteria. We make the Oven-Baked Omelet for Harlots and Ruffians, a recipe we found in a fifteenth-century codex in a library in Milan; we prepare *pomodori bruciati* with our end-of-season tomatoes, an old family dish given to us by a generous butcher in Rome; we roast mussels over an open fire under a pine bough because that is how we saw fishermen do it at a beach on the Adriatic; we cure our neighbor's beef with lashings of salt and pepper to make our own alpine *bresaola*, a cured beef typical of our European counterparts.

In the winery, we follow the simplistic and old-fashioned ways of fermentation for our wine. We choose our harvest date by tasting the grapes; we pick and sort by hand; we crush the grapes with our feet, turning music on loud and dancing, softly treading the fruit. We ferment in open containers covered with clean white cloths. We let the natural temperatures of the season and the climate dictate how quickly or slowly the fermentation takes place. We try to take our cues from the wine, not our egos.

These are very intimate ways of preparing food and wine. We use our hands, our feet, our whole selves in the processes. It is sometimes sublime and often messy. This is why we love what we do: We are engaged with our landscape and try to translate it into something that others can experience along with us. In how I approach our day-to-day struggles and small accomplishments, I have not been unaffected by Caleb's written recipe that begins, "First, I planted the seed . . ." and all that this signifies.

We can all ask why does terroir, or landscape, or a prime ingredient, or organic farming, or natural wine, or native yeast even matter. They must matter. If they do not—if they are relegated to the back of the room, the end of the list, the void—we lose. The elements of living that make our existence profound and worth pursuing are present in such things, and without them, the world in which we live would become void itself.

Fortunately, things like terroir, landscape, ingredients, natural farming, real wine, and native yeasts do matter to many. And they always will matter. Growing food, fermentation, and cooking are the foundations of life; at the most basic level, without them we cannot survive. But they also excite and inspire, sparking our imaginations backward and forward, helping us create lives for ourselves that link us with what is immediately around us and keeping us tethered to the diverse and abundant land on which we love, live, and learn.

Epilogue

When I began writing this book, our farm and winery were really in their infancy. Two years ago I started to formulate my thoughts around how to write about the story of our small farm and vineyard and how we have been learning to farm it. It has been both the worst and the best time to try to write a book about these things. Four years ago, we had just released our first public vintage, and we had no idea how our wines would navigate their world.

WE TOOK SOME COMFORT in knowing that our vineyard was quite small, and the orchard and vegetable gardens all seemed to be relatively manageable for the two of us, along with some very helpful extra hands. Anyone who apprentices with us will tell you that their schedules and their responsibilities are many-faceted, just like how we approach the farming of our land. They must learn the work of the kitchen, the table, and the farm.

In looking forward from that first vintage, we had no idea that the winery would gallop on ahead of us and that we would be flat-out flying behind holding on to the horses' tails. So many changes have happened in those four years it is difficult to catalog them. Those 2010 wines garnered enough interest to encourage us to increase our own plantings on the farm and to look for

other land to plant. We embarked on a roller-coaster ride of partnership and land purchase, investors, grant applications, and business plans. While the initial growth projects did not come together, others quickly replaced them.

This past year, I took over the consulting management of two small vineyards in the Champlain Valley in need of attention, two vineyards where we had been sourcing fruit. The vineyards were owned by two brothers, Bruce and Donald Mina, and sadly in 2010, Donald, the grapegrower, passed away. Bruce, who lives away, inherited these two parcels and tried his best to manage them from afar. Winegrowing is difficult when you are in the vineyards every day, but to try to shepherd a landscape from a distance can be stressful and disheartening. The 2011 season in those vineyards saw sporadic upkeep, a rainy season, and a devastating hurricane. We harvested essentially wild vineyards that made wines that captured the attention of some very supportive wine writers, sommeliers, and palates. In 2012, the two vineyards were managed differently, the larger falling prey to a misguided program of chemical warfare. That vineyard was in such a sorry state come harvest, we could not even pick fruit there. Fortunately, the second vineyard, under different hands, offered up some beautiful berries that again transformed into wine that is now making it into small shops and restaurants in New York and Boston, as well as other places in Vermont besides our farm and the osteria.

In the summer of 2013, a small group of us, including the owner, worked hard at reclaiming the several acres of vine in a season that was defined by consistent rains and pressure from typical diseases and pests. Our challenge—to convert this patchwork growing program to completely organic, to try and rectify the chemical damage in the larger vineyard from the year before, and to do this in one of the most difficult Junes and Julys—was no easy task. The month of August gave us drought, the extreme weather patterns working to our advantage. As expected, the vines that had received solid pruning in the spring had fared the best, but given the circumstances we were all quite happily surprised by how valiant these vineyards are and how much stunning fruit they offered. Cluster quality was not necessarily perfect, but the berries themselves burst with flavor, integrated acidity, and balance. Our surprise was further increased when it became apparent that other producers who had bought significant amounts of fruit in the past chose to decline fruit for 2013 on the grounds that the vineyards had been treated organically. The reasons offered had to do with the use of sulfur and copper in the vineyard as part of the spray program and the concern that sulfurous compounds might be created in the fermentations. Certainly, if you overwhelm your vineyards with sulfur and spray close to harvest, a producer might be at risk for these kinds of aromas and complications. Curiously, no one asked how much sulfur we had used, or when. Given the organic/biodynamic paradigm we follow, and followed for this season, the applications were quite small, even though we had to spray both sulfur and copper more frequently at the beginning of the season due to the heavy rains. Once the weather improved, our spray program tapered off, and more than a month ahead of the harvest we stopped spraying altogether. By the beginning of August the vines were all netted because of the local birds who hunger for juicy grapes, and it was impossible to spray. We were well within the dictates of best management practices. Some of these producers also suggested they would be happy to take fruit the following year if the vineyards were returned to conventional farming.

Suddenly, three weeks before harvest, we were confronted with the opportunity to quadruple our own harvest amounts for la garagista. We dove in and bought as much of the lot as we could. Like so many small wineries before us, we grew the winery in just a few weeks' time with the help of a credit card. In came a new press, a battalion of new demijohns, and

WINE IN THE DEMIJOHN

CLEANING THE DEMIJOHNS

food-grade barrels. Given our cantina, we needed small vessels that two of us could move around by hand both for fermentation and later for elévage.

And there we were, a long harvest covering the whole two months of September and October filled with picking, fruit arriving, destemming, crushing, pigéage, pressing with fruit from both the Champlain Valley and our home farm. Harvest began with serious and dramatic threats of rain, but by the time we arrived in the Champlain Valley, the weather dried out, giving us sun and a perfect breeze, and held with largely warm, Indian summer days and cool, starry nights. Phenomenal volunteers, both locals and folks from down in Boston, moved through our days and nights, doing the constant work that needed to be done. We served food around the clock, and three different guest chefs helped get meals on the table and provided extra dishes that were easy to set up when we needed to break for lunch and dinner. All the while, the osteria was running three nights a week, an amended schedule so that we could manage all the fruit. Breathless and continuing forward headlong, we have not looked back.

While we have *pétillant naturel*, a first-fermentation sparkling wine from 2013, bottled and released alongside our 2012 offerings, the rest of the lot rests quietly in the cantina, coalescing and integrating, the wines' personalities developing, slowly telling us something of what they are and what they want to be. We have since leased the two vineyards in the Champlain Valley, relieving the owner of the stresses of trying to oversee them on his own. Caleb has drawn pages of plans for a new winery at the top of our own vineyard, which we will build in two phases. Groundbreaking has begun. There will be a proper wine cantina next to a root cellar and a place to cure meats. On the main floor, we will have the crush pad, the fermentation and bottling room, and a dry space for labeling and packing.

On the second Sunday of our most recent harvest, two of our friends came and cooked a beautiful Sunday lunch: chicken braised with olives, hummus, beet purée, different salads, and lemon and ginger ices to finish. Bottles of a Corsican rosé wove their way up and down the very long table that was set right in the courtyard in front of the farmhouse. Twenty-two people helped with harvest that day, and we all sat down in the broad sunlight, the sky cerulean blue. The air was warm and it almost felt like summer except for the scent of grape juice in the air. The family of crows swooped down into the rose garden in the vineyard and sat sentinel, watching our antics. Pitchers of newly pressed white juice sat on the table, and we toasted and drank with both the rosé and the fresh juice, honoring the mystery of harvest and the fermentation of grapes into wine.

Someone shared their memory of a harvest they had gone to in Europe, and how like this it was, and how wonderful that they didn't have to travel 2,000 miles to experience it. Our lunch that day made her think she had stepped into one of those French or Italian films where everyone is sitting, talking, gesturing around the table. I laughed and joked that that image was exactly what I was shooting for and trying to capture, and wasn't it a lot of hard work to create a harvest and the need to make wine just so we could all sit down at a long table like this together, sharing in such a romantic yet primordial meal? We all laughed and raised glasses together again.

But my joke really wasn't a joke. At the beginning of this journey, I wrote about the table at the center of the farm, and particularly at the center of a biodynamic farm. And somehow I know that the desire to duplicate these large, raucous communal meals that we had witnessed and participated in so long ago in our life in Italy, hours spent at long, wooden tables—sometimes bare, sometimes with a worn white linen, sometimes covered in a red-and-white-checked oilcloth—encouraged us to seed vegetables, plant a small vineyard, start an orchard, create a bee yard. The image of that table, the

diners all friends, lovers, family, sometimes strangers, where everything else in life is suspended and all that exists at that moment is the wine in the glass, the hot dish being served on the plates, the wooden board of cheese, the half-eaten loaf of bread, a basket of fruit, has led us down this primrose-lined path of tiny kitchens and late nights at our own restaurant. Over the years, the image has slightly changed, or shifted, like one clear shape becoming another, like a pastoral painting of the same view slowly morphing from one season to the next, the tiny kitchen not giving way to, but making room for, a vineyard, a garden, a cantina. There is not only the table surrounded by new and old friends and family laughing and debating, the roast pork or white bean soup, and the little glasses of wine, but there is the phenomenon of sharing in this work together, the work of us within the landscape, this cultivated wilderness, our hands dusty with dirt or wine-stained with the bounty of harvest.

Acknowledgments

AS ALWAYS AT the end of a project like this, there are so many people to thank. The idea for this book had been percolating, but I probably would not have ventured down the road quite yet if it hadn't been for our publisher Margo Baldwin of Chelsea Green. She and Senior Editor Makenna Goodman planted the seed and nurtured the project with patience over these last two years, with gentle prodding when necessary. I thank them for understanding that when the writing wasn't getting done, it was because I was farming, and the book wouldn't exist without the experience of a life living and working with the land. Being reunited with my editor Ben Watson from *Libation: A Bitter Alchemy* couldn't be other than a happy circumstance, and once again I am thankful for his even keel and deep waters of knowledge to help bring the book into a complete shape. I am always thankful for my agent Marian Young, who deals with all the details with such aplomb and has such a keen eye for the big picture. I thank the land on which we live that has taught me so much and has given me the story for this book, and I thank Bruce Mina who owns the vineyards we now lease in the Champlain Valley, giving us the opportunity to grow in a whole new landscape, and Greg Burdick who has been the keystone in helping us do so. To Janey and Taylor and Joseph and Amy who have helped us realize possibilities. The support of friends and family along the way has been so instrumental in keeping the book afloat, and that of the compatriots and volunteers who help the farm and vineyards from bud to harvest like Andrew, our new farm manager who has brought those hoped for chickens and pigs to our land and works garden beds and vines alike. And to my husband, co-conspirator in all things, Caleb, I can never give enough thanks for all the delicious meals, the constant rotation of coffees and teas and fortifying snacks, the unconditional love and support, and for sharing in this adventure that is our small farm and vineyard. Without him, none of this story would have ever happened.

Index

About the Author

DEIRDRE HEEKIN is the author of *An Unlikely Vineyard*. She is the proprietor and wine director of Osteria Pane e Salute, an acclaimed restaurant and wine bar in Woodstock, Vermont. Heekin and her husband and head chef, Caleb Barber, are the authors of *In Late Winter We Ate Pears* (Chelsea Green, 2009), and she is also the author of *Libation: A Bitter Alchemy* (Chelsea Green, 2009) and *Pane e Salute* (Invisible Cities Press, 2002). Heekin and her husband live on a small farm in Vermont, where they grow both the vegetables for their restaurant and natural wines and ciders for their la garagista label.